American Science
in the
Age of Jackson

George H. Daniels

American Science
in the
Age of Jackson

COLUMBIA UNIVERSITY PRESS

New York and London   1968

Preface

This book is intended as a study of key issues in the intellectual history of the American scientific community during the first half of the nineteenth century. It has nothing to do with Andrew Jackson and only indirectly does it have to do with "Jacksonian Democracy." "Age of Jackson" refers only to the fact that the kind of thinking about science described herein reached a height in the 1830s and early 1840s—the period usually given that title. While I hope that the work can be used by others—or perhaps even by me at a later time—as a beginning point for a study of the broad social relationships of science during that period, and also for an exploration of the substantive developments in science, I have limited myself largely to a description and analysis of what the scientists (and "friends of science") were thinking about science. This I regard as necessary groundwork for further studies.

Naturally I have become indebted to a great many people in the course of writing this book. At the University of Iowa, where the manuscript was written as a dissertation, Stow Persons worked very hard to keep my historical sense intact, and Gustav Bergmann was always ready to point out my philosophical errors and suggest fruitful lines for speculation. John Greene of the University of Kansas read an earlier version of the manuscript and offered valuable suggestions, many of which I have incorporated in the present version. By footnotes in the text I have made partial acknowledgment of my indebtedness to him. Robert R. Dykstra of the University of Nebraska gave generously of his editorial and stylistic skill. Bernard Gronert and Frederick Nicklaus at Columbia University Press have been helpful far beyond reasonable expectation. It is a great deal more than a literary convention when I point out that none of these gentlemen can be blamed for my errors, for I have obstinately disregarded their advice on a number of points.

Chapter III is a revised form of an article which previously appeared in the *Huntington Library Quarterly*, and a portion of Chapter VII was published in the *Bulletin of the History of Medicine*. I am grateful to the editors for permission to use this material.

A Woodrow Wilson dissertation fellowship made it possible for me to do the research for the dissertation, and a faculty research grant from Northwestern University was an indispensable aid to me in putting the book in its present form.

GEORGE H. DANIELS

July 1967

Contents

American Science
in the
Age of Jackson

Introduction

There are two fundamentally different ways of approaching the history of science—or, for that matter, the history of any subject whatever. One can either assume that the phenomenon he is studying is a part of a continuing development, or he can assume that it is part of a spatially and temporally bound cultural situation. Few would deny that every human activity has both these aspects at the same time; the only questions usually at issue have to do with significance and possibility. Questions frequently raised are: Which aspect of the subject is the most significant? Is it possible to study the past outside of a developmental framework? An eminent American physicist, crowning a lifetime of speculation about his own science by pronouncing upon a broader subject, answered the first of these questions in favor of development and the second in the negative. According to P. W. Bridgman:

it seems that much of history is not written with an adequate appreciation that the past has meaning only in terms of the present. The impartial recovery of the past, uncontaminated by the influence of the present, is held up as a professional ideal, and a criterion of technical competence is the degree to which this ideal is reached. The ideal is, I believe, impossible of attainment, and cannot even be formulated without involvement with meaningless verbalisms.[1]

One cannot, Bridgman thought, even begin to sever his connections with the present in order to understand the past on its own terms. Many historians, and in particular, historians of science, have at least implicitly agreed with him on this point.

At first glance one can see good reasons for historians of science to choose the developmental approach, either on grounds of possibility or of significance. It is difficult to overlook the huge body of knowledge existing at the present time, and it is equally difficult to escape passing judgment on the past in terms of that knowledge. Science is surely the most progressive subject known—by its very nature it involves a continual building upon the knowledge of the past—and its spectacular achievements in the twentieth century could hardly be denied by either its friends or its enemies. It was only natural, therefore, when historians became interested in science, that they should emphasize this aspect of it. Consequently, we find that, with a few exceptions, histories of science are written within a tightly patterned developmental framework, and all that does not fit that framework is rejected as being of no consequence, or at the very least, of no interest.[2] The historian, having a knowledge of the present state of the sciences, wishes to explain how they came to that elevated state, and to celebrate the achievements of those men of genius responsible for the successive links in the chain leading ever upward to the present. M. Pattison Muir, in the introduction to his *History of Chemical Theories and Laws,* summed up the viewpoint of this type of historian when he said, "The book is not an attempt to move through the past without knowing whereto the course of the science is tending." This approach, he noted, made his work rather "a consideration of the growth of the science from an outside position than an attempt to live through each period along with those who made the advances in that period."[3]

Although most historians of science are still framing their work according to the concept of progressive development, an increasing number have come to realize that this is a terribly limited view of their subject. It has been argued that a too exclusive concern with "scientific progress" tends to obscure the fact that the introduction of a new scientific theory is a profoundly revolutionary act—an act which subverts the entire structure of pre-existing science and is therefore undertaken only in desperation, after all other efforts have failed.[4]

This argument, which I accept as being an essentially sound outline of the way scientific change occurs, suggests some profound consequences for the general historian of ideas. For such a historian, whose interest is in the relations between ideas in different disciplines at a given time, a recital of the successive "links in the chain of progress" is not only useless but positively misleading. This is so because those men who belong in the chain have earned their place there precisely because they were *not* representative of their own generation. They were, for one reason or another, able to transcend the limitations of the scientific community to which they belonged, and to see the world in an essentially new way or they would not have been able to create the innovation for which they are remembered. The study of genius is always fascinating, but such studies reveal only indirectly anything about the normal scientific community of a period.[5]

However useful for some purposes it may be to consider the history of science from the viewpoint of the present, for the purposes of the historian of ideas it is the "normal science" of a generation that is its most relevant aspect. Considered in this manner, with no reference to the chain of progress, science can best be defined as a body of knowledge and opinions about nature, existing at a particular time and place. It is the currently accepted way of framing questions about nature, and the prescribed way of looking for the answers. It even dictates the type of questions that will be asked in the first place, and it determines at what point the scientist will be satisfied with the answers. It is, in short, both a methodology and a general frame of reference, and as such it varies from time to time and from place to place. In this respect, the concept of a "national science" makes perfectly good sense, for the normal science of a generation—like its art, music, or religion—is a part of the larger cultural context within which it appears. It influences, and is influenced by, that context.[6]

The present study is an attempt to accomplish exactly what Bridgman declared to be impossible; to recapture a small part of the past as uncontaminated as possible by present judgments as to importance. It is an effort to understand the thought about science of a generation of "normal scientists"—including practitioners, theorists, and interested commentators. The generation which flourished from the close of the War of 1812 to the mid-century has been chosen as the special object of study. Somewhere around 1815, Americans became interested in the pursuit of science to a greater extent than ever before. They founded journals, organized societies, appealed to the federal government for aid—mostly without success —began sending their students to the scientific institutions of France and Germany in unprecedented numbers, and began bitterly resenting the scientific superiority of Europe. This intensification of activity and the consequent availability of nationally distributed media made possible for the first time the development of an "American scientific community." By the mid-1840s the major problems of theoretical science of the early nineteenth century had been well formulated and new subjects of controversy were beginning to appear. Scientists began speaking less and less of the subjects that they found important in the 1820s and 1830s. Evolution and thermodynamics were on the horizon as new synthesizing constructs. The transportation and communication revolutions had made possible a more effective organization which had profound effects on the life of science in America. Consequently, I have ended my study about 1845.

Although I did not wish to be limited to the writings of any specified, mechanically selected group of scientists, it still seemed desirable to have some objective foundation for the study. Consequently, an effort was made to select a group of men who could be considered "representative" of the normal scientific community of the period. In making the selection I have equated "quantity of work published in nationally distributed journals" with "representativeness." As crude as this method may be, I think there are some rather good arguments that can be urged in its favor. One has at least reasonable grounds for believing that those who published so prolifically in the journals of the profession were the most highly regarded by their contemporaries, and certainly one has presumptive evidence that they were the most read. Since I have tried to define something that could conceivably be called a "national scientific community," I have not included any work published in purely local journals in determining who was a "leading scientist." [7] One argument that can be urged in favor of the resulting list is that it omits only one man who has been identified by historians as having done significant scientific work in America during that period. The individual whose name does not appear is William Beaumont, who did important work on the chemistry of digestion in the 1820s and 1830s. His omission, however, confirms the method of selection, for the fact is that his work was so opposed to the leading philosophical assumptions of American physiologists that it was ignored in this country, and its results not assimilated into American physiology until much later. He was not, therefore, a member of the scientific community of that period.

The only way to decide which journals are of more than local interest is to inspect them all to see if people outside the area of publication characteristically wrote for them. If they did, one has prima-facie evidence that it was read outside the area of publication and could not be called "merely local." Again, if a journal is mentioned frequently in other publications, this can be considered the same kind of evidence. Since there are no reliable circulation data, one has no choice but to rely upon this indirect method. Because of the necessarily impressionistic element in the selection, another list of the sixteen most important journals of the time might be somewhat different; I do not believe, however, that it would be markedly so.[8]

The first chapter is primarily an account of the work of those practicing scientists whose publications in the sixteen above-mentioned journals earned them the title "leading scientists." There were fifty-six of these men, and together they contributed more than half the material published in the journals I have selected during the thirty-year period from 1815 to 1845. The second chapter considers the broader topic of efforts to establish a secure place in American society for the scientific profession. The remaining portion of the work is an analysis of the complex of ideas constituting the "normal" scientific outlook of that period.

After the first chapter the original list of fifty-six men serves only as a control group for testing generalizations. Any general statement, such as "it was generally agreed that . . . ," indicates the following: (1) A systematic study of the writings of the fifty-six leading scientists indicates that the statement is true for them, and (2) I have encountered no reason for believing it was not true of the other scientists of the period. In the first case my information is complete, and can therefore be called "objective"; in the second it is incomplete and must therefore be termed "impressionistic."

In order to avoid making the text both interminable and unreadable, I have resorted throughout the study to the use of lengthy endnotes for three different purposes: (1) to bring in whatever comparative data seemed necessary to substantiate certain points, (2) to make explicit the philosophical basis for critical portions of the work, and (3) to define terms.

CHAPTER I

༄ঋ

The Pursuit of Science

in America, 1815–1845

In 1885 Thomas Henry Huxley looked back over the achievements of science after the second quarter of the nineteenth century and found ample cause for satisfaction. It had been the most fruitful sixty years in the entire history of science, he told his associates at the Royal Society:

the greater part of the vast body of knowledge which constitutes the modern science of physics, chemistry, biology and geology has been acquired, and the widest generalizations therefrom have been deduced, and, furthermore, the majority of those applications of scientific knowledge to practical ends which have brought about the most striking differences between our present civilization and that of antiquity have been made within that period of time.[1]

In contrast to the British scientist's pride in the progress of science in general, consider Simon Newcomb's evaluation of science in America during the first part of that most active period. For half a century after the death of Franklin and Rittenhouse, "there was nothing worthy of the name of national science," he said,

nothing on which the public could look, and say with pride that it was a product of our educational system, or of our effort to promote the knowledge of nature. Two or three men of genius arose, but they received no stimulus to exertion from the public, or, if they did, their works betray the want of attrition with other men of like pursuits able to criticise them.[2]

Both statements seem, from the perspective of the present, to be approximately true, and it is this fact that makes the study of science in America during the first half of the nineteenth century of such interest. During that time England had produced Herschel, Lyell, Darwin, and Joule; France could boast of the completion of Lavoisier's chemical revolution, of the revolution in medicine brought about by Pierre Louis and the Paris clinical school, of the fundamental contributions to electricity of Ampere, and of a host of pioneers in various other fields of science; and German science was represented by the lasting achievements of Schwann, Meyer, Liebig, and Von Helmholtz. One could go on through the list—Norway, Sweden, Denmark, Italy, and Switzerland had produced scientists of international stature. But nineteenth-century America had no one to place in such eminent company before the time of Willard Gibbs. America had seemed to be lagging behind in science, work done here was largely derivative, and at the beginning of the century it was virtually impossible to arouse either public or private support for any scientific enterprise.

Although it was nearing Newcomb's time before the results became obvious, this generally unpromising picture began to change during the early years of the nineteenth century; about 1815, Americans began developing an interest in science—an interest that has continued unabated to the present time, and which made it possible for Newcomb to take a somewhat more optimistic view of the state of American science in his own day than he had of that in the early part of the century.

The period covered by this study—roughly the thirty years after the close of the War of 1812—may be described as the time when American science got its start. During that period American scientists evolved from a disorganized group of amateurs without common goals or direction into the professional body that they had become by mid-century. It is, perhaps, the concern with getting started—with doing the practical things like justifying themselves to the public, organizing institutions, founding media, describing the natural history objects that were being brought from the interior of the country faster than scientific names could be given to them—in short, the prosaic things that had to be done in the early period of a professional community's existence—that best explain the difference between Huxley's statement and that of Newcomb.

Science does not rest solely upon the works of genius; however desirable such efforts may be, several other elements are required to give it a stable foundation. Most obviously, it is difficult for a scientific culture to exist without a particular body of men dedicated to its advancement, and for the successful pursuit of their work these men require some kind of formalized institutional framework. There must be recognized organizations for integrating studies and giving them common direction, an established community for judging, and media for disseminating the results of these studies, institutions for transmitting the body of knowledge so obtained, and it is no less necessary that means of support for those engaged in the work be provided. Before 1815 the United States had been notably lacking in all of these areas. As Governor DeWitt Clinton observed in 1814, the "enterprising spirit," which he thought was a distinct feature of the American national character, "has exhibited itself in every shape except that of a marked devotion to the interests of science." [3] Within the next ten years, however, some evidence of a "marked devotion to the interests of science" became apparent. And as countless witnesses testify, the changes that took place during the next decade after Clinton's pessimistic observations were as obvious to contemporary observers as they are to the present-day historian.

The zoologist James E. DeKay, speaking before the Lyceum of Natural History of New York in 1826, assured his colleagues that great changes had occurred in the condition of the sciences in their country, and he attributed most of them to the successful conclusion of the War of 1812. Previous to that time, he observed, a few isolated works of merit had appeared, but there had been lacking the "familiar interchange of opinions" which was an indispensable condition for scientific productivity. DeKay attributed the early failure to establish a national scientific community to the same general causes as Newcomb cited a half century later. The country was large and its population scattered; transportation was difficult and communication both slow and uncertain; and besides this, Americans had of necessity been too much concerned with founding a nation and settling its territory to pay any considerable attention to science. In short, a nation in the process of establishing itself could neither find much use for, nor could afford to support, a community of men dedicated to scientific knowledge. But he noted that after the end of the war there had been a general awakening of interest in science. A "spirit of inquiry" had arisen among the American people, he said, and they had made notable progress in the natural history investigation of their country. Geology, zoology, botany, and mineralogy—all these had been subjects of a much more intensive interest than ever before.[4]

However much they might complain about the slow start and the persistent failure to match European scientific productivity, American writers were never willing to attribute American deficiency in science to any basic inferiority or ineptness on the part of their countrymen. The typical attitude was probably expressed by the theologian and scientific popularizer Edward Hitchcock, who, in answer to an attack prompted by one of his reviews, conceded that he regarded Europe, not America, as the center of geological study. It was not, however, because his country did not offer so good a field for geological study—as a matter of fact, it was a decidedly better one, Hitchcock thought. Nor was it because European scientists were superior to Americans either in industry or intelligence. It was simply that they had been studying their rocks longer, and in Europe there was both more patronage available for scientific work and less urgent demand for talent in other departments. All that the Americans needed for overtaking Europe was a longer history and a bit more leisure.[5]

The same men who spoke most frequently of the difficulties attending scientific study in America were among those who were quick to point out the peculiar excellencies of their country, and the compensating advantages it offered to scientists. The most common claim was that the relative isolation of their country, and the absence of "unworthy national prejudices" enabled Americans to be eclectic in the best sense of the word; that is, they were able to judge rival European systems with impartiality and extract the best from each of them. The final paragraph of DeKay's *Address* is a good example. He had previously enumerated all the standard reasons for the backwardness of American science in some areas, and he had deplored the lack of support given scientific enterprise by the government, but he pointed out that from another point of view the situation here, removed as it was from European "rivalries and contentions," offered some striking advantages:

we are enabled to examine controverted points with coolness and impartiality. The remoteness of our situation supplies the time and place, and we may be supposed to decide between the conflicting opinions of European naturalists, with the same justice and impartiality as if we were removed from them by intervening centuries.[6]

A few years earlier the medical journalist and educator John Eberle had made the same comment in connection with medical practice in America. "This hemisphere," Eberle said, "is the theatre, on which the prejudices and errors of the European schools, in a great variety of instances, have been refuted and abandoned."

There was, indeed, some justification for the viewpoint shared by DeKay, Eberle, and many others; an eclecticism and a refusal to accept any one exclusive doctrine did seem to be peculiar marks of American science. Eberle gave a specific example in his own field of medicine. The English doctrines of adhesion were fully understood by his countrymen as they were not understood by the French; on the other hand, the improvements of French surgery, which were entirely neglected by the English, were well-known in America.[7] There were examples in other sciences which he could have mentioned; in geology, for instance, most Americans refused to commit themselves to either of the major exclusive schools; in mineralogy one could point to Cleaveland's *Elementary Treatise* (1816), which had drawn equally from the German school and the French school. Cleaveland explained his theoretical impartiality in a preface to the work:

Many of the writers of the French and German schools appear to have indulged an undue attachment to their favorite and peculiar system, and have hereby been prevented from receiving mutual benefit; the one being unwilling to adopt what is really excellent in the other. But it is believed, that the more valuable parts of the two systems may be incorporated, or, in other words, that the peculiar descriptive language of the one may, in a certain degree, be united to the accurate and scientific arrangement of the other.[8]

Although theoretical noncommitment was the more frequently mentioned, there was yet another, and perhaps in the long run more important, sense in which the claim that America was a testing ground for European theories was true. The natural history exploration of the American interior, which began on a grand scale in the first decade of the nineteenth century, had proved damaging to a number of exclusive theories. Classification systems were being rewritten to account for newly described American species, and the geology of America was proving a great embarrassment to both major geological schools. When William Maclure's *Observations on the Geology of the United States* (1817) was published, it had been called upon by geologists of both the Neptunist and the Vulcanist (Wernerian and Huttonian) schools to confirm their favorite theories. A writer in the *Edinburgh Review,* for example, attempted to deduce an argument for the Huttonians from "two remarkable peculiarities" of the geology of the United States as reported by Maclure. First, he noted the unusually rare occurrence of trap rock and porphyries in the eastern part of the United States, and he pointed out that these rocks were almost invariably accompanied in Europe by a dislocation and confusion of the adjoining strata. But the other peculiarity of the geology of the eastern part of the United States, on the Huttonian theory, could perfectly explain this absence of trap rock and porphyries. There was no other place in the world where the same series of rocks extended so far without undergoing any change in the uniformity of their composition and without any disturbance in the regularity of their stratification. This conjunction of peculiarities, he thought, "must be considered an argument of considerable weight in support of that theory of the origin of the trap rocks, which supposes them to have been ejected from below, and to have broken up and insinuated themselves among the superincumbent strata"—that is to say, it was an argument in favor of the Huttonian theory.[9] The reviewer, however, did not comment upon the fact that Maclure "had never seen or heard of any case of limestone, covered by, or alternating with, any rock resembling basalt." Neither did he record the fact that Cleaveland, whose work he was also reviewing, thought the occurrence of basalt in the United States to be "doubtful." [10] As the absence of trap rock in the Eastern United States had posed problems for the Wernerians, the long failure to discover basalt had been a great embarrassment to the Huttonians, and they seized eagerly upon every reported claim of its discovery. By the mid-1820s, however, it was generally recognized that there were facts reported by Maclure which were simply irreconcilable with either school of thought.[11]

It was probably the peculiar features of the geology of the American continent that were responsible for the general marked agnosticism about exclusive geological theories. The wide expanse of the country and the contrasts it offered caused most American scientists to steer a middle course in their theoretical commitments on the subject. Edward Hitchcock, Benjamin Silliman, and Amos Eaton all began their geological careers as Wernerians—as did virtually every early American geologist—but they all abandoned at least some parts of the doctrine quite early. First Hitchcock, and then Silliman, eventually drifted far enough away from their earlier Wernerian commitments to adopt the theory of a central fire. Although Eaton never went this far, he did implicitly acknowledge the fallibility of a central Wernerian doctrine—dating strata by mineral contents.[12] And even those who did not recant were nevertheless quite cautious in applying the doctrine. As the *Edinburgh Review* writer observed of Maclure, "He is a disciple of Werner; but we recognize him as such, more by the descriptive language he employs, than by his theoretical opinions." His general views, the writer continued, "are much more enlarged and philosophical, than is usually met with in geologists of that school; and like most of those who have had opportunities of extensive observation, he has found that the theory of the Freyberg professor is of a very limited application." [13]

In the later stage of geological debate, when the argument between the supporters of fire and the partisans of water had given way to another theoretical division, the typical American attitude was well-stated by the inventor and science teacher, John Locke, in an introductory lecture given in Cincinnati. Speaking of the rival doctrines of catastrophism and uniformitarianism, which were generally thought to be mutually exclusive, he granted that extensive and repeated catastrophes had probably overwhelmed the earth and left their records in its structure; but at the same time he felt that "the powerful agencies now in constant operation," should be given more credit than was customary. "Nor is it possible, perhaps," he concluded, pursuing the via media to its logical conclusion, "to separate entirely the effects of those different classes of causes." [14]

While Americans, in arguing for the potential excellence of science in their country, did tend to emphasize American isolation and noninvolvement with European system-building and national scientific rivalries, they also appealed to the free institutions and breadth of opportunity which American scientists enjoyed. Maclure, writing to Benjamin Silliman from Madrid, noted that the vast advantages attached to freedom were unknown in Europe, and only a few were beginning to understand "the spirit of energy with which a free people pursue whatever they perceive to be for their interest." [15] This kind of argument was much less important than one would at first be inclined to think; perhaps because of general recognition of the other side of the coin—that because of the need for some kind of patronage basic research was a great deal more difficult in such a country.

Although everyone who commented on the subject perceived the need for certain improvements to be made, the level of optimism among American scientists was generally quite high. Of course there remained some who, like James Rush, were convinced that a democratic society was incapable of appreciating the importance of research in basic science.[16] But Rush's viewpoint was definitely a minority opinion.

The prospects of those who wished to pursue science seemed to have improved to such an extent that as early as 1821 Silliman proposed calling his time the *"intellectual age of the world."* It was truly a remarkable time, he thought; for he knew of no other period when the mind of man had been directed toward "so many and so useful researches." In every department of human knowledge one could find talented and industrious men "vigorously engaged in pushing its interests and extending its boundaries, while the press is prolific, beyond all former example, in productions upon every art and every science." [17]

Even in 1821, when Silliman wrote the above comments, the changes since 1815 had been monumental; it is not surprising that he should have been optimistic and touched with pride in his own times. Before 1815 there had existed two general scientific societies in America—the American Philosophical Society (founded 1769) and the American Academy of Arts and Sciences (founded 1784)— but the first was in practice confined largely to Philadelphia in the early years and the second to Boston. One state, Connecticut, had an active academy of science which had been founded in 1799 and had published its first papers in 1810. Only one scientific journal with any claims to national significance was in existence, Samuel L. Mitchill's *Medical Repository.* The first extensive geological study of the United States (1817), the first American mineralogy textbook (1816), the first original chemistry textbook by an American author (1819), and the first native effort at systematic arrangement of American mammals (1825) were yet to be published.[18] When Silliman had been appointed professor of chemistry at Yale in 1802 there were twenty-one other full-time scientific jobs available in the

United States; at the time he wrote his optimistic comments nineteen years later there were apparently more jobs than qualified personnel. Sixteen new colleges had been founded during the time since Silliman was appointed to his chair, and twenty-three additional ones were founded in the decade after he wrote his comments.[19] All of these made some effort to offer instruction in natural science.

By 1821 a large number of scientific societies and "Lyceums of Natural History" were in operation. There were five natural-history societies incorporated in the state of New York alone between 1815 and 1819. In 1819 the legislature of New York provided state funds for the partial support of county agricultural societies, which would have as one of their purposes the general promotion of science. The New York law granted 10,000 dollars annually to be distributed among the different counties in proportion to their population; the proviso was that an agricultural society be formed in each county, the members of which should raise by voluntary subscription an amount equal to the sum apportioned. A further sum of 1,000 dollars per annum was granted to a central board for distribution of seed and printing transactions. New York's early effort at "matching funds" was apparently quite successful, for within one year after passage of the law twenty-six such societies had qualified for their share of the funds.[20] This was a modest beginning, but nevertheless, it placed New York in the forefront of a trend that was nationwide. The rapid multiplication of new societies continued unabated until 1825, after which time it slowed down greatly but never died out. Including all organizations which had as one of their aims the general promotion of science, there were more than three times as many in 1825 as had existed in 1815. The largest increase, reflecting the aid granted by New York, was in the Middle Atlantic Division, where the number quadrupled.[21]

During the same period scientific journals began to flourish, for there was a state of mutual dependence between the scientific society and the scientific journal—even when they were not officially connected with each other. As an inducement for members to participate, the societies needed organs for the publication of papers read at meetings, and the journals needed a ready source of material which, in the absence of any research institutions, only the scientific societies could supply. The earliest scientific publications quite frequently served as quasi-official organs for a number of societies; H. Bigelow's *American Monthly Magazine and Critical Review,* for example, regularly printed between 1817 and 1819 the transactions of the Literary and Philosophical Society of New York, the New York Medical Society, the New York Historical Society, and the New York Lyceum of Natural History. The number of periodicals publishing scientific material more than doubled between 1815, when there were eleven other than the transactions of medical societies, and 1825, when twenty-four such journals were in publication. The centers of publication were New York, Pennsylvania, and, to a lesser extent, Massachusetts and Ohio. However, before 1850 twenty-three states had been the home of at least one scientific periodical.[22]

Scientific journals in that period had an exceedingly high mortality rate for a variety of reasons. Partly, it was simply a sharing of the fate of the entire periodical press, and can therefore be attributed to general conditions. The printing industry, for example, was in a precarious situation all over the country outside Philadelphia, and a share of the printer's and publisher's misfortunes descended as a matter of course upon their publications. In 1836 Daniel Drake reported that it seemed a matter of astonishment that he had been able to keep his journal in publication so long. He had had nine different publishers before the completion of the ninth volume, and had his "editorial vitality not been truly feline" he was sure that his *Western Journal* would surely have long since been defunct.[23] His journal's vitality, however, was apparently no more than "feline," for it did not survive the next change of publishers.

The failure of many early scientific journals was due to quarrels among either their supporters or their financial backers. Speaking of his suspension of publication between January and June, 1839, Drake explained that quarrels in the Louisville Medical Institute had frightened his publishers into withdrawing support until the matter was settled.[24] Somewhat earlier, Robert Adrian's *Mathematical Diary* had gone out of publication because of a quarrel among the mathematicians who supported it.[25] Difficulties like these seemed endemic in scientific journalism, but they became particularly acute with the flourishing of sectarian medical movements beginning in the late 1820s. From this time until the end of the period it was not at all unusual for more periodicals to cease publication in one year than the number of new ones to be founded.[26]

The editors of one of the journals that began during this important decade, that of the Academy of Natural Sciences of Philadelphia, announced in the introduction to the first number what was to be the major program of American science for the next few years:

> In so doing [publishing], they propose to exclude entirely all papers of mere theory,—to confine their communications as much as possible to facts—and by abridging papers too long for publication in their original state, to present the facts thus published, clothed in as few words as are consistent with perspicuous demonstration.[27]

One could not afford to waste more words than were necessary for "perspicuous demonstration," when the flora, fauna, and minerals of America were, in any scientific sense, virtually unknown. Although botany, largely because of its medical connections, had an early beginning, the first American paper on systematic zoology had not been published until 1794, when William D. Peck's description of a new species of fish appeared in the *Transactions of the American Philosophical Society*.[28] In 1817 there were hundreds of thousands of specimens yet to be described, and organizations like the Philadelphia Academy and the New York Lyceum of Natural History were founded to promote their speedy descriptions. In 1820 the American Antiquarian Society was founded as a focus and publication center for the study of a different natural history object—early man in America and the artifacts he had left.

Although special-interest publications began as early as 1804 with George Baron's *Mathematical Correspondent*, the medical journals still had an important part to play. The *Medical Repository*, which under the editorship of Mitchill had brought Lavoisier's chemical revolution to America and had led in the fight against Priestly's effort to save the phlogiston theory,[29] stopped publication in 1824, but more than forty new medical journals had been founded during the decade to take its place, including the *American Medical Review and Journal* (1824), forerunner of the still-existing *Journal of the American Medical Association*. Along with the descriptions of cases and reviews of books, which were their main functions, these journals carried articles on natural history, chemistry, and mechanics, as well as on the expected anatomy and physiology. And unlike the publications of natural history societies, the medical journals were not so likely to exclude papers of "mere theory." They are, in fact, a major source for the student interested in the more theoretical aspects of early nineteenth-century science.

Although no less than 213 different medical journals and the transactions of thirty-four medical societies had been published in America prior to 1850, a relatively small group of men edited the most important of them. The story of American medical journalism during this period could be well told through the lives of John Eberle (who was connected with six different journals), John D. Godman, John Bell, and Isaac Hays. These four men were collectively responsible for the editorship of seventeen different medical periodicals.[30] Editorship of two or three medical journals, given the high mortality rate of such publications in that period, was more often the rule than the exception.

Besides the regular medical journals and general scientific publications, countless special-interest journals were founded during the period, but they were, without exception, unable to survive. Archibald Bruce's *Mineralogical Journal* founded in 1810, could not make it through the first volume, and even in 1842 Benjamin Peirce's *Cambridge Miscellany of Mathematics, Physics and Astronomy* lasted only one year. Similarly, Robert Adrian tried on five different occasions to found a mathematical journal, but only one, the *Mathematical Diary*, published from 1825 to 1832, was in existence for more than a few issues. It appears that there was still not a large enough market for any but a more general publication. A part of the mathematical journal's difficulties may perhaps be attributed to the primitive state of mathematics in America at the time, for it must be said that those that were published contributed virtually nothing to the stock of mathematical knowledge. Number four of the *Analyst* contained Adrian's investigations of least squares, and the *Mathematical Correspondent* contained an essay by Adrian which was the first attempt to introduce the study of Diophantine analysis in America, but for the most part they continued the tradition of the Almanacs—proposing Mathematical "problems" and publishing solutions.[31]

All of these journals played some part in the growth of scientific activity in America, but the crowning achievement of the first two decades of the century, in terms of publication, was the founding of the *American Journal of Science and Arts* by Benjamin Silliman in

1818. The earliest scientific journal in America which has had continuous existence since its inception, Silliman's quarterly could very well be credited with having been the greatest single influence in the development of an American scientific community.

Silliman's periodical, like all the others, passed through a number of difficulties during its early years, which can be followed in part by means of the prefaces the editor wrote to a number of volumes; but through Silliman's efforts and the support of the Connecticut Academy of Science, it managed to weather each storm. Silliman had the usual difficulty in collecting for subscriptions, and much to the embarrassment of Yale College, which offered only moral support for his publication venture, he was forced several times to appeal to the public for aid. By 1829, however, the fortunes of the journal had so much improved that editor Silliman was able to announce that he had begun paying for original contributions. Those who were still willing to write for the journal "from other motives" were requested to do so.

In addition to his financial difficulties, Silliman had to contend with rivals who sought to lure his supporters away, and he also had difficulty with supporters who wanted him to make the journal "more miscellaneous," after the fashion of British publications; but even though the *and Arts* was not dropped from the title during his lifetime, Silliman was resolved to keep it as much as possible a purely scientific journal, for he was "not inclined to renew the abortive experiment, to please everybody." He had included some "miscellaneous" articles in the past, and he would continue to do so, Silliman explained, but in a smaller proportion than was common with British journals.[32]

The pages of the most successful of the early periodicals [33] have provided most of the information on which this study is based. The men who formed what I call "the American scientific community" published most of their work in the sixteen journals that could be identified as "national." In all, 635 men and three women published 2,225 original articles in these journals during the thirty-year period beginning in 1815. Well over half of them (369) wrote only one article each, and a relatively small group of fifty-six men wrote over half the articles (1,116). What the work was like, and what the men who did the bulk of it were like, is of special interest here.

The fifty-six men, as a group, were in the forefront of scientific research and writing among the second generation of American scientists. Their median age in 1830 was thirty-seven; twenty-three of the scientists were born in the single decade 1790–1799. Only four in the group were born before the American Revolution, and only nine were foreign born. Thirty-one members of the group were born in the three states of New York, Pennsylvania, and Massachusetts, the three centers of science and learning in America through the first half of the nineteenth century.

Since no quantitative analysis has ever been made of the scientific work of that period, or of the men who did it, the rather crude data collected during this study may be able to shed some new light on the nature of American science. One finds in the secondary literature a number of rather broad generalizations about the condition of American science during the early nineteenth century—even though no investigation has been made that could justify any of them, and no one has even claimed to have established one of them. One finds, in particular, four points that seem to be generally agreed upon:

1. Natural history, as opposed to the physical sciences, was the dominant research interest and, as such, accounted for the overwhelming bulk of American scientific work.
2. Science was dominated by interest in the practical, as opposed to the merely theoretical.
3. There was a marked absence of specialization during the first half of the century.
4. Science was still largely a pursuit of amateurs, whose main interests, or sources of livelihood, were elsewhere.[34]

While no single study, in itself, is likely to completely overturn any of these generations, just as none has established them, the results of this study, nevertheless, indicate that all must be drastically modified to fit the facts.

First of all, in terms of the total body of work surveyed, there appears to be no foundation for the belief that Americans were uninterested in the physical sciences at that time, or even that they were predominately interested in natural history. According to Table I, the same number of men were involved in publication in physical science areas as in natural history. In terms of the work itself (Table II), almost half of the articles were in natural history, but physical science accounted for only 10 per cent less.[35] The breakdown by five-year periods indicates a steadily growing interest in physical sciences, reaching its height according to Table I from 1840 to 1844, and according to Table II from 1835 to 1839, but at no time was interest in the physical sciences negligible, in terms of either criteria.

While one has only to glance at either Table I or Table II in order to be convinced of the existence of significant interest in the physical sciences in early nineteenth-century America, the question of whether research interests were dominated by considerations of practicality is a great deal more difficult to answer. To settle the question with any finality would require more information than these or any of the other tables contain. One can say, however, that the contention gains absolutely no support from this study, and there is enough evidence to indicate that it is erroneous.

In the first place, if the driving interest of American scientists were practical we would have the right to expect that the bulk of scientific work would be done in those fields of most immediate real or assumed utility. This does not, however, prove to be the case. Botany, for example, was thought to be a useful science because of its connection with medicine. And to make it even more likely that this science would be widely studied if the interests were practical, the great majority of American scientists at that time were trained as physicians. Yet only slightly more than 9 per cent of the 638 scientists published an article on botany, and the work in botany accounted for only 6 per cent of the total publications. Zoology, the practical implications of which were a great deal more obscure, attracted the attention of 15 per cent of the scientists and accounted for almost 17 per cent of the total work. Geology, the most theoretical of the natural-history sciences, attracted the most scientists, while its more practical relative, mineralogy, received no more attention than did botany.

James E. DeKay, surveyng the state of the sciences in 1826, reached conclusions opposite to mine on the relative attention paid to botany and zoology. But the reasons he gave for their relative importance are, nevertheless, useful for this analysis. Writing three years after Torrey's introduction of the natural system into the United States, he remarked that botany had attracted more attention because of "the greater perfection displayed in the systematic arrangement, the physiological history of plants, to the superior attractions displayed by the varied and beautiful forms of the vegetable kingdom, or to the facility with which extensive collections may be arranged and preserved." The neglect of zoology, he ascribed to "the unjust views which have been taken of its relative importance, and to a want of concert in nomenclature and systematic arrangement between the laborers in the different subdivisions of this science." In a "note" to his address, DeKay explained the reason for his own interest in zoology—he thought that the natural system of classification at last offered a chance for men to reconstruct the great chain of being. Artificial systems, still the vogue in botany despite Torrey's efforts, had left the great chain lying "in an apparently confused heap." [36] Although the construction of such a chain could have served no immediate utility, it seems to have been the aim of most American zoologists.

The failure of both federal and state governments to finance scientific research that had no immediate utility has often been cited as evidence that American scientists were practically oriented. Examples of such utilitarian motivations were the early exploring expeditions and the state geological surveys, all of which had to be justified in terms of serving public purposes; contribution to basic knowledge was considered a definitely secondary aim, as a recent student of the geological surveys has argued.[37] But this statement should be more qualified than is customary and applied only to those granting money for research. This failure to distinguish between "research interests of scientists," and "interests of those financing the work" is particularly evident in the frequently encountered allegation that geology was a practical science. Those applying for state aid did invariably argue for the utility of their work, but regardless of this concession to the prejudices of state legislators, there are absolutely no grounds which can be found in the scientific literature that the scientists themselves actually shared these particular preconceptions about the value of their work. On the contrary, it is clear both from their statements and from the kinds of work they elected to do that American scientists were motivated chiefly by what is considered the primary aim of science—the advancement of knowledge without regard for utility. Although Edward Hitchcock, as one historian says, was careful in his state geological reports to give primary attention to economic aspects, it is likely that he conceived it as his duty in those cases because of the source of his

Table I

Men Involved in Research in Various Fields, 1815–1844, and by Five-Year Periods *

Science	1815–1844		1815–1819		1820–1824		1825–1829		1830–1834		1835–1839		1840–1844	
	No.	%	No.	%	No.	%	No.	%	No.	%	No.	%	No.	%
Physics-chemistry	175	27.43	12	17.64	47	24.00	40	23.66	43	25.74	42	23.59	43	26.87
Geology	129	20.21	15	22.05	42	21.53	36	21.30	28	16.76	32	17.97	26	16.25
Zoology	96	15.04	10	14.70	26	13.33	22	13.01	27	16.16	29	16.29	32	20.00
Meteorology	69	10.81	4	5.88	16	8.20	10	5.91	12	7.18	25	14.04	18	11.25
Physiology	67	10.50	8	11.76	22	11.28	33	19.52	6	3.59	4	2.24	5	3.12
Botany	60	9.40	8	11.76	16	8.20	19	11.24	12	7.18	17	9.55	8	5.00
Mineralogy	60	9.40	6	8.82	33	16.92	21	12.42	7	4.19	4	2.24	9	5.62
Astronomy	47	7.36	4	5.88	3	1.53	7	4.14	11	6.58	19	10.67	10	6.25
Paleontology	45	7.05	3	4.41	18	9.23	12	7.10	7	4.19	9	5.05	16	10.00
Mathematics	22	3.44	1	1.47	2	1.03	6	3.55	9	5.49	1	—	6	3.33
Anthropology	16	2.50	1	1.47	5	2.56	1	—	1	—	3	1.13	1	—
Miscellaneous natural history	19	2.97	1	1.47	6	3.07	9	5.32	0	0	1	—	3	1.87
Mechanics	23	3.59	0	0	5	2.56	3	1.77	8	4.79	4	2.24	9	5.62
Physical science total	292	45.76	21	30.88	68	34.87	61	36.09	73	43.71	81	45.50	87	54.37
Natural history total	292	45.76	34	50.00	96	49.23	87	51.47	71	42.51	83	46.62	77	48.12

* Since no workable distinction could be made between physics and chemistry during that period, they are included under one heading. The reason for the blurring of the lines is that the new sciences of electricity and thermodynamics, although generally regarded as belonging to chemistry, were as often as not approached as problems of "natural philosophy." Mechanics, however, is listed separately. The word "physics," of course, was not in general use at that time.

Table II

Articles Published in the Various Sciences, 1815–1844, and by Five-Year Periods

Science	1815–1844 No.	%	1815–1819 No.	%	1820–1824 No.	%	1825–1829 No.	%	1830–1834 No.	%	1835–1839 No.	%	1840–1844 No.	%
Physics-chemistry	551	24.73	21	13.61	143	26.98	104	22.60	99	28.69	102	26.15	82	23.56
Zoology	376	16.87	46	30.26	82	15.47	70	15.21	40	11.59	61	15.64	77	22.12
Geology	270	12.11	18	11.84	71	13.39	61	13.26	47	13.62	41	10.51	32	9.19
Mineralogy	150	6.73	6	3.94	65	12.26	35	7.60	10	2.89	16	4.10	18	5.17
Meteorology	138	6.19	5	3.28	20	3.77	17	3.68	23	6.65	38	9.74	35	10.05
Botany	134	6.01	16	10.52	31	5.84	33	7.17	16	4.63	23	5.89	15	4.31
Paleontology	117	5.25	3	1.97	27	5.09	19	4.13	13	3.76	19	4.87	36	10.34
Physiology	115	5.16	8	5.26	36	6.79	50	10.86	6	1.73	6	1.22	13	3.73
Astronomy	89	3.99	11	7.23	3	—	8	1.73	13	3.76	40	10.26	14	4.02
Mathematics	58	2.60	1	—	6	1.12	8	1.73	26	7.52	6	1.22	11	3.16
Miscellaneous natural history	55	2.46	4	2.54	19	3.59	11	2.39	6	1.73	8	2.05	7	2.01
Mechanics	31	1.39	0	0	4	—	15	3.26	3	—	7	1.79	2	—
Anthropology	25	1.12	1	—	9	1.69	2	—	5	1.45	5	1.28	3	—
Physical science total	867	38.91	38	25.00	176	33.20	152	33.04	164	47.53	191	48.97	144	41.37
Natural history total	1102	49.46	93	61.18	295	55.66	229	49.78	132	38.26	168	43.07	185	59.16
Total	2225		152		530		460		345		390		348	

funds, and because of the explicit instructions that came with those funds.[38] At any rate, the great work of his life, aside from his natural theology, was the collection, description, and classification of fossil bird tracks, a pursuit that could hardly be called narrowly utilitarian.[39]

What the scientists actually hoped to achieve from the allegedly utilitarian surveys was very clearly expressed by one of the most active of the survey workers, James Hall, New York paleontologist. In a long letter to Alexander Dallas Bache dated January 14, 1852, Hall explained that the major difficulty in geology was that geologists had no established points to take as basic for their researches. New York geologists, he said, had taken advantage of the survey for canal routes to fix the elevation of certain formations along the way, but this information had proved inadequate for "accurate geological investigations." After they had ascertained the actual elevation of enough formations, which he hoped would be the result of the next survey, geologists would then be in a position to determine

whether what are apparently parallel and conformable deposits, are really so or not, and then to learn positively what changes were taking place in the bed of the ocean during these apparently consecutive and continuous formations. This point, taken in connexion with the fauna of the successive periods, would be a subject of very great interest. We wish also to know, whether the force that has apparently uplifted all our formations, has acted equally and simultaneously over the whole, or whether some parts of our sedimentary deposits may have, in the course of uplifting, suffered undulatory movements in the time of strike. This knowledge would I am sure, give us a clue to the explanation of many physical phenomena which have occurred since the elevation of our strata, and which in the absence of this knowledge, we fail to understand and properly appreciate.[40]

The striking thing in this list of aims is the total absence of any practical considerations. Clearly, James Hall at least was using the governmental activities in the best way he could to further scientific knowledge.

With regard to the federal government, the case for utility is even less clear. For one thing, as A. Hunter Dupree has shown, the question of what kind of scientific enterprise the government would support was inextricably bound up with the constitutional controversy over "internal improvements" of all kinds—a controversy that was a major political issue.[41] In other words, the decision whether to support a given scientific enterprise was rarely made in terms of *any* judgment about science; it was made, rather, on the basis of one's position on the Constitution or on his sectional political interests. As for any other "internal improvement," it was easier to show constitutional legitimacy for a project which would serve some immediate public purpose—such as an exploring expedition which could be justified either as a military necessity or as an aid to commerce.

But it is by no means clear that those who urged military and commercial reasons for aiding science were actually doing more than finding a convenient constitutional justification for a project formulated for entirely different reasons. The background of the Lewis and Clark expedition is illustrative of this point. In the fall of 1802 Thomas Jefferson asked the Spanish minister "in a frank and confidential tone" whether Spain would object to a group of travelers exploring the Missouri River. These travelers, he said, would really "have no other view than the advancement of geography," but he would "give it the denomination of mercantile, inasmuch as only in this way would Congress have the power of voting the necessary funds." It was not possible, he explained, for the Congress to appropriate funds for a "purely literary expedition," since the Constitution gives Congress no such power. But when Jefferson sent a secret message to Congress in January, 1803, to ask for money for the expedition, his chief selling point was that the expedition would pave the way for wresting the Indian trade in the upper Missouri from the British. He reassured Congress that Spain and France would consider the expedition a mere "literary pursuit." [42]

There is no reason for assuming that Jefferson entirely misrepresented his motives either to the Spanish minister or to Congress. Actually, the expedition would serve both purposes, and Jefferson undoubtedly thought them both important. However, in view of his constitutional scruples, he had no choice but to convince himself (and Congress) that the expedition could be justified by reference to some clause in the Constitution. Jefferson demonstrated his personal interest in the scientific aspects of the expedition by the care with which he had Meriwether Lewis trained as a natural-history collector before sending him into the interior. One important reason for his interest in exploring the continent was his belief in the fullness of the creation and the consequent impossibility of any created species becoming extinct. In the hitherto unexplored areas of the continent he hoped to find living specimens of the fossil forms that had been discovered.[43]

Regardless of the justifications given for such expeditions, all of the government-financed explorers were carefully instructed to bring back botanical and zoological specimens, as well as the obviously useful mineralogical ones—and they were further instructed to bring back not only military and commercial information regarding the Indians, but also to study their languages, their religion, and their manners. The Long expedition, which left from Pittsburgh in the spring of 1819 to explore the region as far west as the Rocky Mountains, was provided with a botanist, a zoologist, a geologist, a "naturalist," and a cartographer.[44] And it was the plant specimens collected on this expedition by Edwin James that John Torrey used as a basis for introducing the natural system of classification to America.[45]

The record seems to indicate that utility has been overstressed in regard to American work in natural history, and turning to work done in physics and chemistry (Table III), one finds that, once again, there is no evident relationship to practical interests. For example, analytical chemistry may or may not be immediately useful, but at any rate, it is the type of chemistry that is most likely to be so. Investigations of the imponderable agents, on the other hand, subserved chiefly theoretical ends. Slightly more work was done on the imponderables than was done in chemical analysis. Organic chemistry, a known useful pursuit, but one for which there was no satisfactory theoretical basis at the time, was generally neglected.

The above comments certainly do not prove that American scientists were not motivated by practical concerns; for it surely might be argued that concentration on areas that had not proved their practicality indicated a desire to discover practical applications for those subjects. There is also, of course, the possibility that scientists simply did not publish their practically oriented work when it failed to have a practical outcome. There are some matters that quantitative data cannot settle, but the least that can be said is that some caution should be used in alleging practical motivations, and the contention will have to be qualified more than has been the case in the past. There is a sense in which this entire work may be considered a nonquantitative argument for the presence of strong theoretical concerns among American scientists; although, of course, theoretical and utilitarian motivations need not be mutually exclusive.

Whatever modern students may have thought about the degree of specialization in the early nineteenth century, scientists of that period were quite sure that their work had already become highly specialized, and that it was becoming increasingly so, although they did not all agree that the trend was desirable. Samuel Jackson, addressing a class at the Philadelphia College of Pharmacy in 1822, pointed to the great advances that had been made in the "mechanic and manufacturing arts" through the division of labor, and called for still more specialization within the field of medicine. Every practitioner, he thought, should have some acquaintance with other fields of medicine, but the subject was really "a universe of sciences—an immense aggregate of subordinate systems" and no individual could conceivably master them all. Given such a condition, more and more specialization was not only desirable but an absolute necessity.[46]

Table III

Breakdown of Articles Written in Physics-Chemistry by Categories, 1815–1844 *

Type	Number	Percentage
Imponderables	204	36.62
Analytical	190	34.11
Gases	62	11.13
Organic	43	7.71
Equipment	35	6.28
Physical chemistry	13	2.33
Miscellaneous	10	1.79
Total	557	

* Six of the articles had to be placed in two categories, therefore, the total does not agree with that given in Table IV. There is no obvious pattern when the work is broken down into five-year periods, and the numbers are too small to be significant.

Benjamin Silliman, however, hoped for an alternative solution rather than for a division of labor within his science of chemistry. He realized that the period had already arrived (in 1839) when no one chemist could master all the facts of his subject, and he granted that "if the full detail of facts" were still to be taught, the science of chemistry would have to be divided under different specialists. He looked forward, though, to the "happy result" when the progress of science would make it possible to present chemistry under a few general principles, "with no more facts than are necessary to illustrate the principles, leaving the immense encyclopedia of the literature of chemistry . . . to be explored as far as occasion may require, without compelling the professor to lead, and the pupil to follow, through every maze of the vast labyrinth." Such was the fate of Silliman's hopes, however, that by 1858 specialization had advanced to such a degree that at least one observer drew up plans for a universal publishing agency to print a "specialist's edition" of separate copies of each memoir issued by all the societies. This, he thought, was a practical necessity if scientists were to keep up with what was happening within their own specialties.[47]

Although specialization in the first few decades of the century had not reached the alarming proportions that Silliman feared, the present survey, nevertheless, indicates that a relatively high degree of specialization in scientific interests was the rule rather than an exception. According to Table IV, of the 269 men who published more than one article in journals during the period, over 40 per cent published work in only one discipline, while another 23.75 per cent published only in very closely related areas. Of the 35.36 per cent who worked in unrelated areas, a large number confined themselves to natural history, and others worked only in the physical sciences. Even if one follows the quite illiberal policy of calling a man "unspecialized" if he published a single article out of his major field of interest, less than 17 per cent can be so designated.[48] Since no study has been made of the interests of either present-day scientists or nineteenth century European scientists, there exists no basis for comparison, but it is clear that specialization was greatly advanced by the time of this study. Table V, which lists specialties and secondary interests of the fifty-six leading scientists of that time, leads to the same conclusion, and it also indicates that some of the sciences were more highly specialized than others. Geology, for example, was a major interest of only three of the fifty-six, but it was a strong secondary interest of fourteen. Mineralogy and meteorology were also more likely to be secondary than primary interests.

Table IV

Writers Publishing Two or More Articles in Scientific Journals, 1815–1844

Type of Publication	Number	Percentage
Publication in one area only	110	40.89
Publication in closely related areas only	64	23.75
Work in unrelated areas	95	35.36
Natural history only	34	12.64
Physical sciences only	16	5.95
Natural history and physical sciences	45	16.73
Total	269	

Table V

Specialties and Secondary Interests of Fifty-Six Leading Scientists, 1815–1844 *

Science	Specialty	Secondary Interest
Physics-chemistry	15	10
Zoology	12	4
Paleontology	7	5
Physiology	4	1
Astronomy	4	2
Mineralogy	4	9
Botany	4	4
Geology	3	14
Meteorology	2	5
Mathematics	1	0
Philology	1	0
Total natural history	30	12
Total physical sciences	22	7

* Under "Specialty" is listed the number of men who did more than half their work in that area. Those who published as many as two articles in a field are defined as having a "secondary interest" in that field. The twelve scientists having a secondary interest in natural-history areas were: four physiologists, two meteorologists, and six chemists. Of the six chemists, five did only natural history work that was related to their major interest. Of the seven having secondary interests in physical science areas, there were three mineralogists, one botanist, one paleontologist, and two geologists. Only one, the paleontologist, did any physical science work that was not closely related to his major natural-history interest.

Finding geology so frequently a secondary interest offers an occasion to comment upon the surprising discovery that very few papers in geology were read before either of the major natural-history societies in America during the mid-1820s, at a time when general interest in that science was at a high point. During the five-year period 1822–1826, geology accounted for less than three per cent of all the papers read before the New York Lyceum of Natural History and the Academy of Natural Sciences of Philadelphia. One could make a good case for the statement that geology appeared to the scientists of that period to be much more closely related to the physical sciences than to natural history. According to the definition of natural history offered by Charles U. Shepard, the leading natural history oriented mineralogist of early nineteenth-century America, geology would certainly not fall within its realm. In a published argument with Benjamin Silliman Jr., Shepard defined natural history as a general department of natural science whose business it was to perform certain services for the other branches of science. It dealt with the natural [physical] properties of objects, providing them with a classification and nomenclature, furnishing a set of identifying characteristics, and providing a full description of the properties. According to Shepard, natural history could have absolutely nothing to do with either composition or historical development—the very subjects that geologists were most concerned with. The function of natural history was merely to describe and classify that which was *here and now*.[49] Shepard's seems to have been the common viewpoint, and this, I think, explains the lack of interest in geology by the natural-history oriented members of the societies. Among those scientists who wrote on geology during the period, it was associated with work in the physical sciences in twenty-three cases, and with natural-history work—predominantly paleontology or mineralogy—in eighteen. The reason for the physical scientists' interest is that speculative theories of geology then in vogue were either astronomical (the nebular hypothesis) or chemical, as is well-illustrated by the case of Lardner Vanuxem, which will be discussed in a later context. In much of Vanuxen's work, it is difficult to decide whether a specific paper should be classified as "chemistry" or "geology," since he used his theories of both to reinforce each other. At any rate, this apparent lack of interest in geology by those who were oriented enough toward natural history to join a society dedicated to its study, accompanied by the interest manifested by physical scientists, reinforces the claim made before that many American scientists had strongly theoretical interests.

As for the last generalization—the amateurishness of scientists—Table VI indicates that it, too, is without foundation in the sweeping form in which it is usually made. Most of the scientists, in fact, would have to be regarded as trained professionals. While it is certainly true that no chemist, for example, had a degree in chemistry (for such degrees were not given), yet it is not true that most of them were untrained in their professions. The fifty-six men who did most of the scientific writing at that time did, in fact, constitute an extraordinarily well-educated group. Almost half of them had medical degrees, which were conferred as a result of actual attendance at a medical school, and twenty others were trained at a college where scientific instruction was offered. Only the remaining nine could, by any stretch of the imagination, be called self-educated in science. Among the twenty-one who had been partly educated in Europe, Paris and Edinburgh seemed to share top popularity.

Table VI

Occupational and Educational Data on Fifty-Six Leading Scientists

Education	*Number*
Medical training in college	27
College other than medical	20
No college training	9
European education	21
Edinburgh	6
Paris	6
England	3
Italy	1
Holland	1

Occupation	
Professor of sciences	41
Other employment as scientists	5
Practicing physicians	3
Other	7

Along with being a remarkably well-educated group, America's most prolific scientists were also a professional group. Thirty-six members of the American Philosophical Society and twenty-seven members of the American Academy of Arts and Sciences were included in the group. Only fifteen did not belong to either of the major scientific societies. Forty-one of the fifty-six were usually employed as professors of sciences in a college or university, while only seven had no scientific employment. Together they occupied sixty different professorships at thirty-six institutions. Five of the seven who could be termed 'amateurs' worked exclusively in natural-history areas.[50]

The overall picture of American scientific activity during the early part of the nineteenth century that arises out of this survey is a very different one than is often drawn. The erroneous notion that historians have of American scientific work is most likely a product of their commitment to the idea that the only relevant aspect of science is its progressive development. This commitment has caused a peculiar blindness to all scientific effort that could not be neatly incorporated into the developing body of science. American natural history observations were readily incorporated, for the natural history exploration of the American continent was a major event in the history of science. North American scientists described new species of plants and animals, described new geological formations, and American physicians isolated new diseases. Whatever the fate of natural-history theories, natural history descriptions are not so likely to be superseded as they are to be elaborated and refined. The history of scientific progress, therefore, has a place for those who wrote the early descriptions. But the work of Americans in the physical sciences was quite generally superseded, and since they made few "contributions" to these sciences, a history of the progress of science fails to indicate that they were vitally interested in this area. The same point could be made concerning the lack of theoretical interests and, indeed, of the other two generalizations as well. It was not that Americans were uninterested in theory; it is, rather, that they were found almost invariably on the side of a theoretical debate that a later generation decided was "wrong." [51] An over-zealous attention to scientific progress has obscured the entire nature of the early nineteenth-century scientific community.

If one attends only to the scientific efforts of early nineteenth-century Americans, we must conclude that there was in this country a flourishing scientific community which, by the early 1820s, had adequate organs for the publication of its researches and the beginnings, at least, of a well-developed institutional structure. More often than not, the scientists had quite specialized interests, and the leaders of the scientific community were a well-educated, professional group of men. From the number of different academic institutions represented by these men (see Appendix I), it appears that opportunities for the study of science existed all over the more settled areas of the country. None of the ordinary reasons given for the lack of scientific achievement in this country during that period seem evidently to fit the facts. Certainly America did not possess the highly developed institutional structure that France had, and certainly it lacked England's long tradition of scientific research.[52] These factors alone would have led one to expect a certain amount of retardation in science. But retardation is one thing; the virtual lack of achievement during that period is quite another. Clearly, American achievements did not even measure up to such potentiality as its trained manpower and the institutional supports it did possess would have led one to expect. At least a part of the blame for this gap between potentiality and achievement must be placed upon far less tangible factors than physical limitations. Such "intangibles" are the subject of the remaining part of this work.

∞∞∞

CHAPTER II

The Scientific Profession

As the gentleman amateur had been the prototype of the man of science in the eighteenth century, by the mid-nineteenth century the trained specialist—the professional whose sole source of support was his scientific employment—had come to be the new type. The emergence of a community of such professionals was the most significant development in nineteenth-century American science. Most of the controversies within science can be understood in terms of this development, and the necessity for a professional body to justify its existence to the public even played a major part in establishing the peculiar philosophy of science shared by the first generation of professionals.

In 1802 there were about twenty-one full-time jobs in science in the United States, all academic positions.[1] This includes those who were teaching scientific subjects, such as chemistry or botany, in medical schools, but it does not, of course, include practicing physicians or teachers of the art of medicine. This picture began to change very quickly after that date, however. By the 1820s the demand for teachers of science had actually outrun the supply. The change is dramatically illustrated by the fact that in 1824 Thomas Jefferson had to send to England for professors for his University of Virginia. Jefferson, who was opposed to importing Europeans for such jobs unless necessary, explained that otherwise he would have had to raid other American schools, for there were no qualified persons available who were not already employed.[2]

The characteristic home of the American scientist in the nineteenth century was the college or university, while in the eighteenth century it had been the private laboratory, or the privately organized week-end natural history outing. Both the college and the college-employed scientist had existed before, but even here the scientist was becoming less and less like his eighteenth-century academic counterpart. The man who provided scientific instruction at the colleges before 1800 had characteristically been a "natural philosopher"; a man with broad interests who taught all the sciences from a philosophical point of view as part of a humane curriculum.[3] Although natural philosophy was early distinguished from natural history, and by about 1800 had begun to be distinguished from chemistry, as an academic position "natural philosophy" included all the sciences and could perhaps most usefully be compared, both in scope and orientation, with the modern academic concept of "general science." The individual scientific specialties developed out of natural philosophy on the one hand and medicine on the other.

The process of differentiation and division of functions, in terms of interest, was based upon the increasing complexity of the sciences and increasing knowledge about nature. During the opening decades of the nineteenth century, at the same time that science began to pass out of the public domain into the hands of professionals, it outgrew the capacity of one man. In terms of specialized jobs it was based upon practical considerations and the necessity of competing for personnel brought about by the rapid growth of older colleges and the creation of new ones. At the smaller colleges in the 1850s one still found professors of natural philosophy who taught all the sciences. At the larger colleges increasing enrollment led to the appointment of additional professors and the division of previously existing chairs. For example, Yale in 1802 had a professor of mathematics and natural philosophy who also taught whatever chemistry and natural history was offered. In that year Benjamin Silliman was appointed to the newly created chair of chemistry, natural history, and mineralogy, and these areas were taken from the incumbent in the chair of mathematics and natural philosophy. Upon Silliman's retirement in 1853 the chair was further subdivided, and his son became professor of chemistry, while James Dwight Dana became professor of natural history. In 1864 a professorship of botany was established and Dana became professor of geology. Meanwhile, the other half of the chair that had been split when Silliman was employed was also going through the same kind of division. The professorship of mathematics and natural philosophy was divided in 1836, with Denison Olmsted retaining only natural philosophy. This process, of which Yale is only one example, continued throughout the century and is, in fact, still continuing. The decentralization and the highly competitive nature of American academic institutions made it inevitable that other colleges would attempt to emulate the standards set by the most advanced.[4]

While it was the increasing complexity of the sciences that made a body of professionals necessary, at the same time it made it increasingly unrealistic to urge the public to understand the work of the scientist in his own terms. The complaint of John Ware, Boston physician, naturalist, and dabbler in chemistry, is symptomatic of the reaction of nonprofessionals to the recent changes in chemistry:

it is difficult to avoid regretting the overthrow of so simple and beautiful a system as that of Lavoisier. It is impossible for those of us, who have formed our ideas of the chemical operations of nature on the principles which he taught, to turn with complacency from a theory like his to a state of science so unsettled and so obscure, as modern chemistry is. . . . They have made it less captivating to the general scholar; they have lessened the interest with which it is viewed by those not immediately engaged in its pursuits, by rendering it more complicated and more difficult to be understood, and less applicable as a whole to the explication of those phenomena of the natural world with which we are most familiar.[5]

Ware, speaking from the point of view of the talented amateur who had suddenly become an "outsider," spoke with praise of the discoveries of professional chemists, yet regretted their tendencies. "Insiders," such as the highly professional Robert Hare, likewise expressed their perplexity, although in a somewhat different tone. In the same year (1819), Hare wrote to his friend Benjamin Silliman with evident irritation: "I was told in New York that many said they could not understand my memoir, who considered their standing such as to feel as if this were an imputation against me rather than themselves." Hare protested that he could not write for those who were so ignorant "without making it too prolix and commonplace for adepts." The scientist's difficulty, he continued, was that "we cannot write anything for the scientific few which will be agreeable to the ignorant many."[6]

Complaints like these were very frequent in the first part of the century—both from "outsiders" like Ware and "insiders" like Robert Hare. For different reasons both were dissatisfied with the changed condition of the pursuit of science since the time of Lavoisier. Their comments amount to saying that chemistry had passed beyond the capacity of the average man—or, more significantly, of the talented amateur. Ware, speaking on behalf of the amateurs, was uneasy about having been passed over; Hare was dissatisfied by the demands made by the amateurs on the professionals, and by their presuming to criticize his work when they plainly could not understand it.

The system of Lavoisier had been a great advance over earlier chemistry; it had rescued chemistry from its affiliations with alchemy and had made it a respectable scientific discipline. But it was, after all, a simple system—in fact, it was its very simplicity that had been the main argument in its behalf. Being based upon the assumption that each element had its characteristic function in any compound—as, for example, oxygen was the "acidifying principle" —and containing rigid dichotomies, such as the division between "combustibles" and "supporters of combustion," it was an easily comprehensible system. With a copy of Lavoisier's table of the elements and knowledge of a few general principles, one could be said to have mastered the science, in some sense. But a few years after Lavoisier's death it was known that nature did not behave so simply as all that. Davy's isolation of the alkali metals, sodium and potassium, in 1807 had practically dispelled the view that certain earthly substances might be elementary; and his demonstration, about four years later, that chlorine was an element, not an oxide, had overthrown Lavoisier's view that oxygen was the exclusive "acidifying principle." Work on gases shortly after the turn of the century by Robert Hare and others had already led to questioning the validity of the distinction between "combustibles" and "supporters of combustion." Worse yet, it was beginning to appear that one had to know not merely one mode of behavior for each element, but how each element reacted with every other element. In a word, chemistry had developed a body of knowledge of such complexity that only a professional group possessing special qualifications could be entrusted with its keeping, for only such a group could be expected to fully comprehend it. The specially qualified group would, as a natural consequence, begin to experience new difficulties in communicating with those outside their profession as soon as their subject became esoteric. It was the dawning awareness of a kind of alienation that Robert Hare was expressing as early as 1819.

Natural history soon went the same route that chemistry had gone. From the viewpoint of the development of the profession, the real significance of the adoption of the natural system of classification in botany and zoology and the chemical classification of minerals was that these innovations created esoteric bodies of knowledge where they had been parts of the common stock of knowledge before. The Linnaean system of classification was based upon obvious, easily grasped physical characteristics, such as the number and position of the sexual parts. The natural system, on the other hand, required the student to compare a large number of relations in order to arrive at a judgment of "overall affinity"—a judgment that could successfully be made only with a complex of knowledge possessed by the trained expert. It was the same with mineralogy. Physical classification since the time of Werner had been arrived at on the basis of lustre, hardness, color, brittleness, and other readily ascertained characteristics. The new system, on the other hand, required a knowledge of chemical analysis. In geology, the crucial event was the increasing use of fossils as guides to dating strata—again, a task requiring specialized knowledge and wide agreement upon principles of nomenclature.

The sum of all these changes was the creation in the early nineteenth century of an esoteric body of knowledge called "science." [7] As one would expect, the term "scientist" was also coined at this time to refer to those who had previously been designated "natural philosophers." William Whewell was the first to use the term, and by the 1840s in America the terms "scientific men" or "natural scientists" were being used to distinguish such people from other kinds of philosophers. Likewise, the term "philosophy" began to take on its modern definition. Professional behavior was defined at this point in terms of the acceptance or rejection of the innovations which were making the subject esoteric. The less professional, such as Amos Eaton, Thomas Nuttall, and Chester Dewey, resisted the trend to the end in botany and zoology; likewise Charles U. Shepard and other like-minded scientists resisted the innovations in mineralogy.

The vehicle through which most of the esoteric elements were introduced was the *American Journal of Science and Arts*. Silliman, the founder and chief editor of the *Journal*, has had his scientific competence much maligned—even by his younger contemporaries —but at this period he would qualify as a champion of the professional interests. Not only did he steadfastly resist efforts to make the *Journal* "more popular" or "more miscellaneous," but he and his junior editors systematically worked to introduce the new views into American chemistry, beginning with the first issue in 1818; geology, beginning in the 1820s; zoology in the 1830s, and botany and mineralogy in the 1840s.

I am not here suggesting that a deliberate conspiracy existed on the part of those arguing in favor of the changes. After all, both innovations were certainly significant advances and they had important, purely *scientific* (as well as professional) causes and consequences. But the important point is that the changes served the described function when viewed in the context of the creation of a professional body of scientists. And, as a matter of fact, participants in the controversy seemed perfectly aware of what they were doing. As Chester Dewey, one of the important opponents of the natural system, said to Asa Gray in 1842, "The natural method takes botany from the multitude, & confines it to the learned." Earlier, Thomas Jefferson, on much the same grounds, had rejected the natural system and expressed his preference for the Linnaean, even though he was not completely satisfied with the latter. The difficulty with Linnaeus' system was that some determinations required anatomical dissection. It would certainly be better, he thought, to base a system upon "such exterior and visible charactes as every traveller is competent to observe, to ascertain and to relate." [8]

Recognition that the natural system confined botany to the learned was usually voiced more explicitly by opponents, for those who were urging the changes naturally did their best to publicly deny that they would have any such consequences. John Lindley, for example, devoted two pages of the preface to both the 1845 and the 1853 editions of his *The Vegetable Kingdom* to arguing that the difficulties of the natural system had been grossly exaggerated.[9] And Asa Gray, in reviewing an earlier work of Lindley's, argued that with the aid of the natural system one could actually obtain "a really useful knowledge of the vegetable kingdom" with less labor than ever before.[10] But, of course, the claim that the natural system was easy to master was an overstatement, if not an outright attempt to deceive; and many scientists did admit that the subject contained obscurities that could not be penetrated by the layman, although they generally tried to point out some compensating advantages from the layman's point of view. In a long article intended to demonstrate the beauties of the natural system, one would-be popularizer, coming to one of the most important parts of the system, hurried over it, noting in passing that he should say a great deal more on that point: "but as it would be difficult to render our language intelligible to any but practical botanists, we are reluctantly obliged to dismiss thus briefly the most interesting among all the discussions on this subject." [11]

It was also during the 1840s that the term "popularizer" began to be used more frequently and to take on its modern invidious connotation. Popular lecturers generally abandoned the old effort to convey up-to-date scientific understanding, contenting themselves instead with retailing "wonders." This appeal, which was virtually absent in the first quarter of the century, had become common by the 1840s—a sure indication of the inability to communicate on other terms.[11a]

The extent to which science had been removed from the nonprofessional understanding is well illustrated in a popular lecture that Edward Hitchcock was giving in the decade before mid-century. Hitchcock, a highly competent scientist in several fields as well as an ordained minister, attempted to illustrate the superiority of the sciences over fiction by making a survey of the various sciences and exhibiting "wonders" from each. His painful awareness that he could not expect audiences to comprehend anything above the level of superficiality was evident throughout the lecture, and in several places he explicitly stated as much. In speaking of mathematics, a subject "full of curious and fascinating truths," he had to admit that it was so profound "that I shall scarcely be able to clothe even one fair example in a popular dress." On this subject he had to content himself with relating a number of "paradoxes" designed to stimulate the appetite for further knowledge. Although he (perhaps wisely) made no effort at all to explain his illustrations, his hearers must have been awed when they were told such things as that a man may approach nearer and nearer to a fixed object eternally, and yet not be able to reach it; that two lines may approach nearer and nearer to each other forever without meeting (the asymptote to the hyperbole), and that one infinitesimal (which is by definition infinitely small) may yet be infinitely smaller than another. "How stupid must that intellect be," Hitchcock explained, "which is not roused and interested by such paradoxes!"

The same realization that he could not really communicate was evident in his discussion of most other subjects. The principles of mechanics "for the reasons mentioned in speaking of mathematics" could not be exhibited; only such minds as Newton, LaGrange, Laplace, and Bowditch could fully understand astronomy; and as for chemistry now that it was based upon the doctrine of definite proportions, "He only can fully appreciate its beauty who has long been devoted to the delicate and difficult department of chemical analysis." [12]

While the removal of a body of knowledge from the public domain is a necessary first step in creating a place for a society of experts, this removal also places special obligations upon those who wish to pursue the knowledge on a professional basis, for in a Democratic society professionals cannot exist without a measure of public support. Since the public can no longer be asked to share, in any real sense, in the inner mysteries of the profession; and since they cannot reasonably be asked to support the joy of discovery or intellectual curiosity of someone else, some other basis of contact must be found. That is to say, the emergent profession must justify its work in terms of its social value. It is incumbent upon the profession to demonstrate that it is in the interest of society at large to support a special group in the cultivation of esoteric knowledge. One obvious method open to the scientist is to appeal to practicality, and in the nineteenth century this was widely used. But in America it took a unique form appropriate to the Jacksonian context in which the appeal was made. The strategy here was to link appeals both to patriotism and the Democratic assumptions of Jacksonian America with appeals to the general utilitarian spirit of the age. As observed in the preceding chapter, even scientists whose private statements and research interests were wholly unconnected with immediate utility were nevertheless anxious to publicly represent themselves as utilitarians. Science was recommended to all who would hear as the highest example of useful knowledge. It was widely proclaimed that it would not only enrich and elevate the individual, but would make the nation great in both peace and war and advance mankind in general on the road to that happiness for which it was destined.

Cultural patriotism, of course, was not a creation of the Jacksonian period—although it reached its highest development at that time. As early as 1802 Samuel L. Mitchill's *Medical Repository* found cause for satisfaction in the fact "that among the graduates of Edinburgh . . . there was not a single one from the United States." To Mitchill this was evidence that American medical schools like the ones at New York, Philadelphia, and Cambridge "are engaged in the business of medical education to an extent that is both pleasing and surprising." [13] According to Benjamin Silliman, writing a few years later, it could be confidently predicted "that until intelligent chemists are trained up at home, and induced to attempt the introduction and extension of the CHEMICAL ARTS, the United States will never attain to the pinnacle of national superiority, which Great Britain and France owe more to the *successful cultivation and application of natural science,* than the one does to the prowess of her armies, or the other to the triumphs of her marine." [14]

After the close of the war of 1812, which was widely regarded as the final achievement of American independence, the patriotic appeals became even more direct. One of the major results of that war, according to James E. DeKay, was that a "proper feeling of nationality" had been widely diffused among American naturalists. This feeling had "impelled them to study and examine for themselves," he continued, "instead of blindly using the eyes of foreign naturalists, or bowing implicitly to the decisions of a foreign bar of criticism." [15]

Another writer, celebrating the natural history exploration of the American continent since the signing of the treaty of Ghent, which he also regarded as the key event in recent history, said that it was one of America's chief national blessings that had resulted from the peaceful condition since that time "that we have been enabled and induced to turn our thoughts *inward*." The spirit of investigation that had developed during that period should be pleasing to the political economist, the moralist and the patriot alike. "Whether we look to its effects on the wealth or on the happiness of our community, we are sure that to cultivate and to cherish it must be regarded as a sacred duty." [16]

The Philadelphia Journal of the Medical and Physical Sciences was founded by Nathaniel Chapman in 1820 as an effort to answer the attack on American culture and science made by Sidney Smith in the *Edinburgh Review*. Smith had written a general indictment of American civilization—of its art, literature, and drama, as well as its science—but Chapman was particularly incensed by the question "What does the world yet owe to American Physicians or Surgeons?" This quotation was carried on the title page of Chapman's journal for almost five years as a deliberate method of urging upon Americans the patriotic duty of supporting an indigenous science. In the prospectus published in the first number, Chapman pointed out that in regard to medicine Smith's charge was as unjust as it was insulting, for the current medical literature of Europe was describing doctrines and modes of practice which had been adopted there after long usage in America. As a matter of editorial policy, Chapman declared that his leading aims would be:

to trace the progress of medicine in the United States, to vindicate our claims to certain improvements, to preserve these, as well as what may hereafter be done, from foreign usurpation, and lastly, to evalue, and stimulate the genius of the country to invigorated efforts.[17]

Chapman's cultural patriotism was also excited by the appointment in 1824 of an Englishman (Robley Dunglison) to the chair of the Institutes and Practice of Medicine at the University of Virginia. Although he was ready to concede that in many instances a man might be imported who was at that moment better qualified than any available American, Chapman saw no reason either for despair or for the wholesale importation of foreigners. There was no doubt, he thought, that the apparent superiority of foreigners was merely a temporary condition arising out of the better opportunities for leisure in Europe. If an American scientist were given such a position, he said, echoing a question asked many times by his contemporaries, "shall we not soon make a man superior to any one who could be brought from abroad?" Chapman found it hard to believe that Thomas Jefferson, the father of the University of Virginia, could have had anything to do with Dunglison's appointment, for it was inconceivable that he would have chosen a foreigner in preference to providing opportunity for an American. In fact, according to Chapman, a favorite aphorism of Jefferson was that "the failure of almost all the great scientific or literary undertakings of Americans, is to be attributed to their employment of foreigners, instead of calling into exercise the talents of their own citizens." [18]

Views like these were expressed by scores of scientists and writers of reviews in the early part of the nineteenth century. J. G. Cogswell, reviewing John W. Webster's geological studies of St. Michael and the Azores, had his pleasure increased by the fact that it was the production of "an author of our own." His national pride, he said, was gratified in seeing such an encroachment upon the domains of the Old World, although he did regret that Webster had been forced to go abroad to get his advanced education. Cogswell, who had himself studied in England and Germany, considered the necessity for going abroad to get an education not only belittling to the homeland but positively dangerous from the national viewpoint. "Genuine, glowing patriotism proceeds in part from prejudices in favor of one's country," he warned, "and these prejudices are very liable to be scattered by the comparisons, which travellers are sometimes obliged to make in foreign lands, to the disadvantage of their own." [19] Even though Cogswell grudgingly admitted that European science was more highly developed than American, and he did inject the warning note, the dominant tone of his article was still one of pride and high hope for the future.

Whenever a reason was given for the backward state of American science a major role was assigned to the necessity of working for a living in a democratic society, and the lack of either private or public patronage for scientific research. The federal government, in particular, came in for criticism because of its niggardliness, and patriotic appeals were often linked with invidious comparisons of American and British governmental support of science—no doubt as an effort to awaken a sense of shame in the public mind. John D. Godman, well-known physiologist and propagandist for science, reviewing the achievements of Major Long's second expedition to the Rocky Mountains, was pleased to note that a great deal had been accomplished with the limited means available to the expedition; but at the same time he said it was a matter of regret that it had been so poorly financed, and that the government had not provided more adequate publication of its findings. Certainly, he thought, England would not have given her scientists such shabby treatment, and he could not understand why America should do so.

It was "mortifying," he said, for an American to witness the "magnanimous policy" of England, and then to contemplate the pitiful contrast exhibited by the United States Government. Britain, Godman warned, was even then sending its explorers to the borders of the United States, while the American government was sitting idly by and taking little interest in the work of its scientists. This "magnanimous policy" of England was also an example of enlightened self-interest, Godman was careful to point out, for it was contributing more than anything else to the wealth and strength of that nation. The Congress, Godman acknowledged, "by some unaccountable generosity," had voted the money to send the expedition to the Rocky Mountains, but by forcing the men to travel against time before the limited funds were exhausted Congress had very nearly destroyed the scientific value of the exploratory trip. And as for the governmental effort to save a few hundred dollars by inadequate publication, Godman pronounced it "of all attempts at economy, the most ill-judged, narrow, and preposterous." [20]

Such invidious comparisons of American with British support of science remained quite common throughout the period. In 1840 an anonymous reviewer in the *New York Review* compared the fortunes of the American, W. C. Redfield, who had provided the best support to date for the rotary theory of storms, and his British follower, Colonel Reid. In the first place, the reviewer maintained, it seemed unjust that the author of such an important theory should be compelled to print his researches at his own expense, or else be confined to the limited space available in the monthly miscellanies. Colonel Reid, who was at the most a "faithful and intelligent follower of Redfield," had been able to publish all his work in "the most splendid style" by the aid of his government. But there was yet another way, the reviewer pointed out, that the British government had demonstrated its high regard for Reid's scientific ability. While Redfield's labors had brought him neither reward nor appreciation from his government, Colonel Reid had been appointed to a sinecure in Bermuda, an ideal place to pursue his researches. Such a contrast, the reviewer said indignantly, was not "creditable to us as a nation." [21]

In addition to pointing out that Redfield had personally been treated unfairly, the reviewer had attempted to appeal to the sense of national pride by suggesting that such treatment threatened to discredit America in the eyes of the world—to deprive it of its rightful place as a great nation. Such an effort to shame Americans into more active support of science was evident in every field of science, but it took an especially prominent place in the continuing pleas for a national observatory. American astronomical work had previously been confined to the limited facilities available in a few universities and small private observatories, but by the 1820s most people understood that the time had passed when isolated individuals could contribute significantly to astronomical knowledge. Because of the need for ever larger and more complicated instruments and for large-scale coordination of efforts, it was thought to be especially fitting for the government to enter this area of science. The technique of argument is well illustrated by this statement in the *American Quarterly Review* for June, 1828:

We fear the time will come, when some English critic, some Quarterly or Edinburgh reviewer, will exhibit to the world the mortifying fact, that for years after every petty German principality had its astronomic institution, supported at public expense; after the date of the establishment of an observatory in the convict land of Botany Bay, the people of the United States were so devoid of scientific intelligence, as to have made no provision for such a purpose.

Rather than wait for this "severe truth" to be urged as a reproach by foreigners, the writer said that he preferred to disclose it himself, "and to call upon those who guide the councils of the nation, to deliver us from the reproach." [22]

The idea of a declaration of scientific independence from Europe remained the most common plea of those who were urging government support of science throughout the period. One of the earliest and most persistent of the many movements for establishment of a national observatory was connected with the need for a prime meridian in the United States. It was widely held that rather than follow the common custom of reckoning from either Greenwich or Paris as zero degrees in determining longitude a sovereign nation should extricate itself from the "degrading and unnecessary dependence on a foreign nation" by laying a foundation for fixing its own first meridian.[23] William Lambert, a government clerk and one of the organizers of the abortive Columbian Institute for the Promotion of Arts and Sciences, concluded that continuous astronomical observations were necessary to provide accurate bases for the required fundamental surveying, and during the first decade of the century he launched a campaign to secure governmental support for such a project. In 1827 he was still arguing on the same grounds for an observatory where "we might observe and compute for ourselves . . . and . . . prepare and publish an Astronomical Ephemeris, independent of the aid of European calculations." [24] Speaking on the same subject, a Whig writer in 1845 argued that an astronomical observatory would enable American navigators to keep their course through the seas "in patriotic reliance upon the calculations of their countrymen," rather than upon those of some foreigner who, presumably, could not be trusted with the welfare of American ships.[25]

It was in this same spirit that Benjamin Silliman, in a "note" to an article on the Coast Survey, said that the suspension of its operations was a great blow to science, but it should also be considered a national misfortune. This was so, he continued, not simply because of the loss of previous expenditures or of the delay in practical benefits—both of which he thought significant arguments—but it was especially unfortunate because the principles that Superintendent Hassler used were in advance of the science of Europe at that period and, consequently, their application to practice originally in the United States would have "redounded to the national honor." [26] But in spite of the persistence of such complaints and invidious comparisons with the British government large-scale support of science by the government was not forthcoming in the nineteenth century.[27]

In explaining themselves to their countrymen American scientists had to stress practical values so much for the very reason that it was not at all obvious how a great deal of their work could be of any utility. The public understood the value of steam engines and other examples of useful technology, but it was not so evident to the popular utilitarian mentality that learning exactly what heat was, or whether electricity was composed of one fluid, two fluids, or none, was useful in the same sense that, say, designing a better safety valve was useful. And in spite of the scientists' protestations of innocence, the suspicion persisted in the public mind that they were in reality more interested in some kind of abstract research—in some kind of theory—than they were in practical applications. This ingrained suspicion of the man of useless knowledge was a characteristic of Americans that some scientists understood very well, and they took special steps to counteract it. Samuel L. Mitchill, in a review of H. M. Brackenridge's *Views of Louisiana* (Pittsburgh, 1814) noted that Brackenridge had expressed his apprehension "lest the character of a man of science be fastened upon him." Editor Mitchill regretted his attitude, but conceded that Brackenridge had sound reasons for it. "In a state of society like ours," Mitchill explained, with just a trace of bitterness, "where a mere suspicion of possessing eminent attainment of this kind, too often lessens the confidence reposed in an individual as a man of business, it is certainly discreet to avoid the imputation." [28]

The American public has always had difficulty recognizing the connection between abstract science and useful technology. According to one disillusioned commentator, a botanist in America was quite likely to be told to "stop gathering good-for-nothing weeds, and take to some honest and profitable employment"—words that one gathers had been applied to the author on some recent botanizing expedition. The public, as the hapless botanist said, was utterly incapable of understanding any object in science except that of direct practical utility.[29] Since it must have been evident to anyone looking over their journals that scientists did not characteristically concern themselves directly with practical things, scientists were careful to stress that the moral and religious aspects of science were as valuable to society as was its practical utility. If they could not appeal to the ignorant part of the public in this manner, scientists could at least make their subject respectable to "thinking and well-educated minds" by calling attention to "the more elevated and philosophical portions" of their work.[30] In a society that could not afford "mere entertainment"—when the dominant literary criticism, for example, frowned darkly upon all kinds of fiction—science could be recommended as the highest kind of "rational amusement." And to a public which believed firmly that nature was the creation of an omnipotent and benevolent God the study of God's works could be represented as a duty, not merely an idle pastime. Unlike the purveyors of fiction and other vanities, scientists dealt with God's own world, and by understanding the natural mechanisms of that world they were contributing to knowledge of God's plans and to knowledge of the moral government of the world. There was, in a word, a priestly side to the work of the scientist. On the basis of the "wonders" he had unfolded in the popular lecture referred to earlier,

Edward Hitchcock had no doubt that "the choice of every noble and ingenuous soul" would be science as a substitute for romance. "Science is the history of nature," he told audiences all over the northeastern United States. It is "the history of the works of the Deity," he continued, "And shall the inventions of man come into competition with the inventions of the Deity?" [31]

In the arguments for teaching science in the schools, moral values were stressed almost to the exclusion of any others. "The proper study of nature begets devout affections," one writer pointed out in arguing for the introduction of that subject as a regular classic into normal schools and academies. This undoubted truth, he continued, had given rise to the common maxim "That a true naturalist cannot be a bad man." [32] A very strong conservative appeal could be made on the basis of the moral values of science, and propagandists did not let this opportunity go by unnoticed. Edward Hitchcock gravely informed the conservative readers of the *North American Review* that knowledge of chemistry and mechanical philosophy should be more widely disseminated among the "less informed classes of mankind," for it would powerfully conduce to benefit their morals.[33] A still stronger statement of the same nature was made by the Unitarian minister and popular scientific writer, W. B. O. Peabody, writing in the middle of the Jacksonian era. In this country he thought it exceedingly important that a taste for such studies should be generally spread, for "we have already too much of that excitement, which grows out of the quick sympathy of party associations." He observed that it used to be thought that Americans were not an easily inflammable people, but the truth was that the means of "igniting them and arranging them in systems" had only recently been discovered—presumably by Andrew Jackson. Now that such means had been discovered science offered a solution.

The best, perhaps the only remedy, is to supply resources, and engage as many minds as possible in pursuits of a less stirring description; science, literature, and all the elegant arts, will be so many calm and unpretending, but still efficient, means for securing the peace and happiness of our nation, which is endangered only from within, and which will die by suicide if it perish, and not from any foreign blow.[34]

In other words, science, literature, and the "elegant arts," being "less stirring" subjects, would keep the democratic horde out of trouble by diverting its attention from politics.

Even though scientists made effective use of the conservative fear of unenlightened masses, when American propagandists for science were enumerating the benefits of their subject they always stressed, in keeping with their democratic context, that the benefits extended to all classes. In his answer to Thomas Carlyle's attack on mechanistic civilization, an American writer (Timothy Walker) suggested that because of the cultivation of science "the terms uphill and downhill are to become obsolete," and that machines would soon be performing all the drudgery of man "while he is to look on in self-complacent ease."[35] Becoming somewhat lyrical in an introduction to a treatise on scientific method, Samuel Tyler in 1844 said he proposed

to give some account of the philosophy of utility—the philosophy of lightening rods, of steam engines, safety lamps, spinning jinnies and cotton gins—the philosophy which has covered the barren hills and the sterile rocks in verdure, and the deserts with fertility—which has clothed the naked, fed the hungry, and healed the sick—the philosophy of peace, which is converting the sword into the pruning hook, and the spear into the ploughshare.[36]

And still another commentator, writing in the *New Englander* in 1851, claimed that by virtue of scientific advances more had been done within the past fifty years than in all previous recorded history to place the comforts and "elegancies" of social life as well as the necessities within the reach of all classes.[37]

The scientists were able, by varying the nature of their appeals to suit the particular audience, to make quite a good case for themselves. The government was told that national honor demanded that science be given support and the public in general was told of the vast practical benefits that science had to offer. The conservative fear of social disorder was effectively exploited by pointing out the quietistic function of science (as well as that of education in general), and the more liberal were offered the prospect of bringing social "elegancies" within the reach of all classes. Once this rather comprehensive argument was developed it is not surprising that the possibilities of science so quickly captured the imagination of Americans, and the verbal—if not yet the pecuniary—support of most public officials.

But with all their success—one might say, largely because of it—one last obstacle remained. Perhaps the most successful of the arguments scientists made was that their subject had important moral implications. At any rate this was the one argument for which little disagreement could be found in the literature of the period. When one considers what normally happens at some point in the process of professionalization, it seems quite surprising that so little dissent could be found to these moral implications, for at times propagandists for science came very near to representing themselves as instructors in morality. In doing this, of course, they opened themselves to the charge that they were arrogating to themselves the prerogatives of another, already well-established profession.

In the process of defining a position for a new profession, a measure of encroachment upon already established professions is almost inevitable; that is to say, the new profession generally claims special competence in some area that has previously been the exclusive domain of another. One necessary part of the job of emerging professionals, then, is to define their relationship to those with whom they are coming into competition. In the nineteenth century (as in the eighteenth) the scientists' chief rivals were theologians, for both groups considered themselves to be dealing with natural laws, and both considered themselves competent interpreters of those laws. In such a situation the surprising thing is not that there has been some "conflict between science and religion," but that there has been so little of it. In early nineteenth-century America there was virtually none, and this lack of conflict can possibly be explained by a peculiarity in the relationship between the two professional groups. The theologians, logical rivals for the scientists, were also the keepers of an important part of the cultural values to which the new profession had to appeal if it were to become firmly established. This meant that any appearance of conflict had to be avoided at all costs—and the burden of avoiding the conflict was on the new professionals. The job of scientist had to be defined so that it appeared to supplement that of theologian, rather than displace it, or even take over any important part of its duties.

How this was best to be accomplished was a problem that had given Edward Hitchcock, a respected member of both professions, a great deal of anxiety. In 1852, speaking at the Andover Theological Seminary, he attempted to spell out a code of ethics to regulate relations between the professions. Each group would have to give up something, but Hitchcock thought a mutual understanding would work to the benefit of both. On the one hand, it was necessary for theologians to give up their absurd efforts to derive scientific principles from the Bible, for it was not a tenable claim that revelation had anticipated scientific discovery; the aim of revelation was essentially different, having to do with salvation and God's overall plan for mankind. The scientist concerning matters of fact should, therefore, be allowed the "freest and the fullest liberty of investigation," for what he discovered could have nothing to do with God's overall plan. Scientists simply discovered *how* God did things—matters that He had not deigned to reveal in the Bible—and they gave the term "law" to His customary mode of operation. On the other hand, scientists must admit that as a means of moral reformation and a regulator of human affairs, their subject had no *direct* power. After these concessions, the only thing remaining necessary was for the two professions mutually to adopt as a maxim the principle that "entire harmony will be the final result of all researches in philosophy and religion." The facts and principles of science, Hitchcock said, "to an unprejudiced, sophisticated mind"; are favorable for piety, for those sciences form a "vast storehouse" for the use of natural theology, and they "cast light upon and illustrate revelation," after it has been received by other means. Hitchcock concluded with a description of his ideal man of science:

He is a man who loves Nature, and with untiring industry endeavors to penetrate her mysteries. With a mind too large for narrow views, too generous and frank for distorting prejudice, and too pure to be the slave of appetite and passion, he calmly surveys the phenomena of nature, to learn from thence the great plan of the universe as it lay originally in the divine mind. Nor does he stop when he has found out the mechanical, chemical, and organic laws of nature, but rises to those higher principles by which the moral relations of man to his Maker are disclosed. Hence he receives with gratitude and joy those richer disclosures of truth which revelation brings. To its authority he bows reverently and rejoicingly, and counts it the best use he can make of science to render it tributary to revelation and to the cultivation of his own piety. He exhibits a generous enthusiasm in the cultivation of science; but he has a stronger desire to have it associated with religion; . . . and we may be sure that whatever goes by the name of science, which contradicts a fair and enlightened exhibition of revealed truth, is only false philosophy.[38]

This paragon described by Hitchcock was able to indirectly exercise the moral function and thereby contribute to the reformation of society simply because he was studying the laws by which God regulated the matter of the physical world and directed it for the benefit of his creatures. Every instance of order scientists found in nature, if properly interpreted by a learned and pious mind, could be used to reinforce belief in a benevolent God, and consequently, to impress upon men the necessity of obedience to His laws. Science, in a word, was to be the true handmaiden of theology—words that as professional a scientist as James Dwight Dana actually used in a public address.[39] The general acceptance of this conception of the scientist's task meant that even the abstract researcher—no matter how far removed he was from steam engines, spinning jinnies, or cotton gins—could be described as performing a service function of great public utility.

Natural theology—the effort to prove the existence and display the attributes of God by evidences from nature—has frequently been commented upon as having been peculiarly evident during the first half of the nineteenth century. Appeals to the evidence of nature had been common long before, but theoretical expositions of the foundation of natural theology with relation to the natural sciences were nineteenth-century phenomena. Despite this fact, scholars have failed to understand that natural theology, in terms of the science of its day, was a perfectly respectable scientific discipline, and furthermore that its chief proponents were scientists. All available evidence indicates that the vogue of natural theology was not a case of theologians misusing science for their own ends, but of scientists trying to attach some of the aura of the theologian to their own profession; that is to say, misusing science for *their* own ends. Theologians, in fact, were a great deal more likely to stress the inadequacy of arguments from nature—although of course they were generally happy to use them for whatever support they offered. And because historians have failed to understand this they have not considered how being primarily a natural theologian whose job was to refine the description of an Evangelical God, affected the scientist's work and his attitude toward it in that formative period. It is impossible to overstress the importance of the pervasive belief that science was to be the handmaiden of theology, for it affected, in one way or another, most of the debates within the scientific community in early nineteenth-century America.

Very early in the century there was some suspicion that science and religion were not the entirely compatible disciplines that later apologists made them out to be. An anonymous pamphleteer, writing an attack on Thomas Jefferson's science in 1804, could not find a good word to say for "philosophers." The writer's comments on Jefferson's account of the Blue Ridge Mountains is characteristic of his treatment:

The first ambuscade of infidelity, according to custom, is among the mountains. Whenever modern philosophers talk about mountains, something impious is likely to be near at hand. Not more numerous are the streams which flow from the Alps and the Andes, than the objections which they have afforded to these sophisters against the sacred history. When mountains are mentioned in their writings, the well-meaning reader has need to guard against some wicked insinuations with as much vigilance as he would against the lurking panther, if he were passing through the forests which shade the sides of those mountains.[40]

In the generation following Jefferson American scientists were as anxious to repudiate the enlightenment philosophers as the anonymous spokesman for religion over science had been. But they, unlike Jefferson's assailant, did not find it necessary to draw any invidious distinction between religion and science. In the brief time since the evangelical revival at the beginning of the century apologists for science and for religion had managed to accommodate their subjects to each other. The extent of their success can be inferred from the fact that such criticisms as that quoted above are extremely difficult to find in the literature of the 1820s and 1830s.

The scientific community profited enormously from its association with conventional religion—however unfortunate for both science and theology this association ultimately proved to be. The common assertion that "it must first be known that there is a revealer before it can be known that anything is revealed" could not fail to elevate the status of the scientist in a society that wanted desperately to know that something had been revealed.[41] Naturally, it was always stressed that only the methods of the natural sciences could provide indisputable proof of a Revealer's existence. As Benjamin Peirce warned possible opponents of science, the church simply could not afford to renounce the external proofs of a God which he as a scientist was more than willing to provide.[42] In a word, the particular form of the association with religion not only gave legitimacy to the pursuit of science but it gave the society at large an important stake in that pursuit. If there were no scientific proof of the existence of

God, it was argued, revelation would be stripped of all authority, "and man must forever live amidst the spectres of uncertainty and spiritual want." To a generation that feared uncertainty above all else nothing could have been more comforting than to assure itself that "no theorem in mathematics is more certain than the existence of God." [43]

A critic of Locke and the Scottish philosophers several years earlier had commented upon the presumed connection between religion and Scottish philosophy, the latter being the vehicle through which Baconianism was diluted before being transported to America. It was the peculiar misfortune of our country, James Marsh thought,

> that while the philosophy of Locke and the Scottish writers has been received in full faith, as the only rational system, and its leading principles especially passed off as unquestionable, the strong attachment to religion, and the fondness for speculation, by both of which we are strongly characterized, have led us to combine and associate these principles, such as they are, with our religious interests and opinions, so variously and so intimately, that by most persons they are considered as necessary parts of the same system; and from being so long contemplated together, the rejection of one seems impossible without doing violence to the other.[44]

Although it is questionable whether matters had reached such a state at the time Marsh wrote, twenty years later his observations would certainly have been true. The scientists by that time had indeed been able to connect their version of "the philosophy of Locke and the Scottish writers" so intimately with the religious interests of Americans that it seemed impossible to attack the one without attacking the other.

After the identification had hardened into an orthodoxy, the theologian Tayler Lewis was placed in exactly the position that Marsh had described earlier. In writing an attack on science he rejected natural theology, and he rejected, as well, any attempt to harmonize science and religion. In a later work Lewis continued his attack on science, arguing that it was both deceptive and futile. Science could not answer the deepest and most perplexing questions that occurred to man as he beheld the wonders of nature. It was, he insisted, totally incompetent to deal with questions of origin, of purpose, and of the essence of things.[45] Tayler Lewis made few converts with his attacks, for his argument was widely held to be not only a libel against science but a profanation of religion. It was the scientist James Dwight Dana who marched most decisively to the defense of orthodoxy; his argument is a monument to the nineteenth-century faith that science and religion could not clash. It is significant that Dana's attack on Lewis, which assumed book-length proportions through four installments, was published in a theological journal—not in the *American Journal of Science,* of which he was a co-editor, but in *Bibliotheca Sacra.* Lewis's argument, Dana held, embodied a "fatal misconception of science," for science had never even claimed to explain the ultimate nature of life—nor was it so "presumptuous" as to hope to explain it. But this should not be considered a failure of science. "Has metaphysicl or sacred science yet explained the nature of God or spirit?" he asked. "Or has any mind yet measured eternity?" Scientists had the humility to realize that no form of investigation could penetrate to such mysteries, he continued, for "the ultimate nature of matter or life is as much beyond all investigation." In this respect scientists had not sinned, Dana reassured his readers, but misguided theologians like Lewis who provided examples of this "prying into mysteries" in their writings on the second revelation.[46]

Dana received such a heartening response to his refutation of Lewis that in the second installment of his argument he felt strong enough to announce a manifesto. "With such an exhibition of the Bible before him," he warned, "its enmity with science insisted upon, if he [the scientist] is not so grounded in faith as to be sure his opponent is wrong in his hostility, he will feel forced to stand by nature, God's acknowledged work, *versus* the Bible, 'the Book.' " [47] Theologians should tread softly, it seemed, if they wanted to maintain the mutually profitable relationship with the scientists.

Once a profession has defined its orthodoxy and has settled upon a way to conduct its relations with significant groups outside the community it must find a way to deal with transgressors from within. Every profession has certain members who either violate its agreed-upon standards or, by refusing to accept the compromises involved in legitimation, threaten its standing in the community. These members must be dealt with if the profession is to survive. Frequently used self-policing methods are licensing regulations, the threat of withdrawing privileges or other formal means; and the profession usually enlists the aid of the state in enforcing its decisions. Scientists, however, can use no such formal measures because of their strong commitment to "openness"; this is regarded as being necessary to ensure the discovery of new knowledge. Nevertheless, it is absolutely essential for the survival of the profession that some means of control be found. The method that they did adopt (and the method still followed) was much more subtle, but just as effective as a licensing system.

A member becomes dangerous when he either violates the methodological canons of the profession or makes some attack on societal values that may cause the public to look askance at the profession as a whole. In order to deal with such an erring member the defenders of orthodoxy try to associate these two sins with each other—that is to say, they try to provide public evidence that in attacking societal values he was, at the same time and of necessity, violating the methodological canons of the profession. In other words, he is not a true representative of the profession, but a charlatan. If it is adopted as a postulate, as Edward Hitchcock suggested, that "entire harmony will be the final result of all researches in philosophy and religion," the assumption that immediately follows is that if entire harmony is not the result in a given case, a methodological error or an error in interpretation must be present as the only acceptable explanation for the apparent lack of harmony. In providing the necessary public evidence it is important that the offender be dealt with by a recognized spokesman for the profession. This explains the fervent denunciations of anything approaching evolution theory during the early part of the nineteenth century, for the striking thing about these denunciations is that they usually came from scientists, not from theologians.

Robert Chambers' *Vestiges of Creation,* published anonymously in 1844, is a case in point. Of all the bad reviews this book and its evolutionary theory received, the most intemperate came from scientists. For example, the burden of Asa Gray's fifty-two page review in the *North American Review* was that the principles used by the author of *Vestiges* were not in keeping with the best scientific thought; if they had been the book could not possibly have contained all the religious heresies found in it.[48] Two years after publication an anonymous writer in the *New Englander* summarized the earlier comments on *Vestiges'* author: "He may indeed be a bitter foe of our holy religion, but we are more ready to accuse him of attempting to seduce science from her legitimate sphere." [49] Throughout his generally temperate review the writer took pains to point out to his readers that it was well understood that Chambers' errors were not errors of science but merely of inaccurate or outmoded science. This is an important claim to make, for it not only relieves the profession in general of the individual's guilt, but it also becomes possible to assert that maintaining true religion depends upon maintaining true science—thus giving the public another stake in preserving the scientific orthodoxy. Despite the fact that private statements by several leading American scientists indicate that they favored the views put forth in the book, it had no American defender. In Henry D. Rogers' opinion, expressed privately to his brother, it contained the "loftiest speculative views in Astronomy and Geology and Natural History, and singularly accords with views sketched by me at times in my lectures." [50] Rogers, however, did not review the book.

But the furor with which scientists greeted the same author's *Explanations* (a sequel to *Vestiges*) was a great deal more intense than was the reception of the first book. Asa Gray, who wrote forty-one pages this time, was highly irritated. The cause of his perturbation became evident on the next to the last page of the article, where he quoted Chambers' comments on his scientific reviewers. Chambers, while acknowledging that scientists were almost universally against him, did not regard their judgment as conclusive. ". . . The position of scientific men with regard to the rest of the public is such, that they are rather eager to repudiate than to embrace general views, seeing how unpopular these usually are." Going further, Chambers had suggested that scientists were so engaged in their own little departments of science that they failed to see the big picture.[51] Gray protested:

Any such man, deeply versed in a single department, is much better qualified to judge of the whole scheme, than one who, like our author, professes to possess only a superficial acquaintance with any branch of science whatever. Who but the men of research have ever established sound and comprehensive views of nature or have made stable generalizations in any branch of science?

Gray summed up the case against the author of *Vestiges* as follows:

When, therefore, the unprofessional reader of the *Vestiges* learns that Brewster condemns its physics, Murchison and Lyell and Sedgwick its geology, that Cuvier (in advance) as well as Professor Owen, Agassiz, and others, scouts its comparative anatomy and physiology, Whewell its whole logic and philosophy, and every sound-minded man we have yet heard of, its theology and tendency,—why, we ask, should not the unprofessional reader rely upon their independent testimony, in respect to facts which they are the most competent witnesses of, and inferences of which they have the best means of judging? [52]

Gray was angry, in short, because the author had refused to submit to the judgment of the scientific community and had tried to take his case to the public. He pointed out that Chambers' speculations, "if we may judge from the general testimony of scientific men," had been repeatedly weighed in the balance and found wanting.[53] Once this had been done, Chambers' insistence upon appealing to the public over the censure of the profession was a cardinal sin. Gray was always particularly sensitive about the unique competence of the scientific community to judge of scientific matters. As he wrote to J. D. Hooker of Agassiz: "A has a touch of the empiric about him, in that he is always writing and talking *ad populum*—fond of addressing himself to an incompetent tribunal." [54]

Even though Agassiz himself was under suspicion by some members of the scientific community for his frequent appeals to an "incompetent tribunal," he was as eager as Gray to condemn the practice in others. He also condemned the "unscientific parentage" of Chambers' theories, and said that the author of *Vestiges* "casts a stone at our Scientific Institutions" with his demand for a jury of ordinary readers.[55] Hugh Miller, in his book-length case study intended to refute the *Vestiges,* stated the case that scientists were anxious to make quite simply: "All the leading assertors of the development hypothesis have been bad geologists." [56]

In addition to castigating him before the public and disclaiming any professional responsibility for his errors, one effective way to punish an erring member was to deny him access to journals. Because a small group of men edited the most important journals this could be done quite effectively. Denying a scientist the right to publish is a more than symbolic excommunication from the profession, and the step was taken a number of times during the early part of the century. Silliman announced as a definite policy that while he was soliciting communications on all scientific subjects he would reserve the liberty "to return those which are not adapted to his views." [57] As long as Silliman's views coincided with those of most American scientists, as they did early in the century at least, his editorship of the most important scientific journal was a definite asset to the community—despite the frequently voiced disparaging opinions of his scientific ability.[58]

In keeping with this policy, a sheaf of papers was returned to C. S. Rafinesque in 1819, and he was never again able to publish his researches except in his own fugitive journals and privately printed pamphlets. But despite the fact that Rafinesque had been virtually ostracized from the profession for more than twenty years, upon his death it was still necessary to make the ritualistic public disclaimer. The ceremony was performed by Asa Gray, who wrote probably the most remarkable obituary in the history of science a few months after Rafinesque's death. Gray began the notice in the tone of one who is performing an unpleasant but necessary duty. In the early days, Gray explained, Rafinesque was in some respects greatly in advance of the other writers on the botany of the United States, but "we are obliged, at the same time, to protest against all of his later and one of his earlier botanical works." Gray's reason for writing the rather lengthy notice, he said, was because ". . . from the manner of their publication, many of his scattered writings are little known to men of science." Here is the rather remarkable spectacle of an exceedingly overworked scientist taking time out to write a careful consideration of works that he considers entirely worthless, and justifying it on the grounds that the works were "little known to men of science." Gray certainly needed to flog no dead horses in order to enhance his own reputation at this time, and without a knowledge of the ritualistic value of the condemnation, his writing such a notice would be inexplicable. Gray's notice did serve an important purpose, however, for Rafinesque had broached a particularly offensive form of the evolutionary hypothesis, which Gray stated as follows:

According to his principles, this business of establishing new genera and species will be endless; for he insists, in his later works particularly, that both new species and new genera are continually produced by the deviation of existing forms, which at length give rise to new species, if the foliage only is changed, and new genera when the floral organs are affected. He assumes thirty to one hundred years as the average time required for the production of new species, and five hundred to one thousand years for a new genus.

A radical evolutionism like that of Rafinesque would be a double danger in that it would not only violate the general cultural norms but would also make a chaos of Gray's science of botany; it was therefore necessary as a kind of self-reassurance to demonstrate that Rafinesque was not a competent scientist. His "passion" for establishing new genera and species was said to have become a "monomania," countless stories of his hasty classification were detailed, and, with a triumphant flourish, Gray told the story of Rafinesque sending a paper to the editor of a scientific journal "describing and characterizing, in natural history style, *twelve new species of thunder and lightening!*" Surely, Gray was reassuring his readers— as well as himself—one need pay no heed to such a man.[59]

Partly on the strength of this carefully preserved harmony between religion and science—and, no doubt, in part because of the other reasons discussed—interest and faith in science very quickly came to be practically universal among literate Americans. One could read praises of science in popular review organs, in newspapers, in scientific journals, and in Protestant Church periodicals.[60] In fact, an American writer who wanted to defend mechanical philosophy in 1831 had to turn to England in order to find a worthy opponent. Timothy Walker, writing in the conservative *North American Review*, composed an eloquent answer to an article by Thomas Carlyle, who had complained that even poetry, the workings of genius itself, was no longer without a scientific exposition.[61] The American champion of science saw no reason why it should not be so. "Does the poet merely rave?" he asked rhetorically. "Is his mind lawless in its wanderings?" Unlike Carlyle, his critic could see no injury that mechanism had done to man, although he conceded that the "presumptious intermeddler" had taken some liberties with nature:

Where she denied us rivers, Mechanism has supplied them. Where she left our planet uncomfortably rough, Mechanism has applied the roller. Where her mountains have been found in the way, Mechanism has boldly leveled or cut through them. Even the ocean, by which she thought to have parted her quarrelsome children, Mechanism has encouraged them to step across. As if her earth were not good enough for wheels, Mechanism travels it upon iron pathways. Her ores, which she locked up in her secret vaults, Mechanism has dared to rifle and distribute. Still further encroachments are threatened.[62]

Walker's was, perhaps, an extreme statement, and a great deal of it might have been deliberate hyperbole, but it was still only a few degrees removed from the norm. Other periods in history have been noted for an enthusiastic confidence in science, but there had been none before in which the feeling was so widely shared.

CHAPTER III

◈

The Reign of Bacon

in America

Edward Everett, editor of the *North American Review,* Unitarian minister, and Massachusetts politician, began a review in 1823 with a remark that might very well characterize the intellectual temper of the period in which he lived. "At the present day, as is well known," he observed, "the *Baconian* philosophy has become synonymous with the *true* philosophy." Everett's choice of the adjective "true" was not a matter of accident—it was not merely that Francis Bacon's philosophy was the most adequate or the most useful, but it was thought to be *true,* and any other philosophy was correspondingly *false.* The philosophy of Bacon was so familiar to all the readers of the *Review,* however, that Everett thought it quite unnecessary to speak further of it. Instead, he would confine his remarks on the book under review to the section having to do with Bacon's political career; a subject on which, presumably, he might be able to enlighten his readers.[1]

The Baconian philosophy so dominated that whole generation of American scientists that it is difficult to find any writer during the early part of the nineteenth century who did not assume, with Everett, that his readers knew all about it. Dugald Stewart, in his history of philosophy, also disclaimed any need to speak of Bacon's *experimental* philosophy on the grounds that this was so well known as to be obvious. In his chapter on Bacon, the Scottish philosopher considered only his ethical and political philosophy and his philosophy of mind.[2] It had become fashionable—indeed practically universal—since the reaction against the Enlightenment to begin any scientific treatise with a paean to the "Baconian method." Sometimes instead of Bacon the name of Newton was used, but in either case the meaning was the same, for the thinkers of that period made no distinction between the scientific approaches of the two men. Bacon had urged that scientists concentrate on "fact," and Newton had said he made no hypotheses; Newton's work was cited by countless writers as an example of pure Baconianism, and Bacon was just as often credited with having been a precursor of Newton, or even of having anticipated him in making his most important discoveries.[3] But for the most part, no one would speak of either beyond acknowledging their leadership. Exactly where these two dignitaries were leading nineteenth-century men of science generally remained a matter for inference. A quotation from John Esten Cooke, a popular medical writer of the period, may serve as an example of the bow scientists usually felt called upon to make:

Long since convinced that experiments and observations are the only true foundation of knowledge, and that hypothesis is the ignis fatuus by which we are led astray, the author of the following pages has endeavored, in the investigation of the changes produced in the system by the remote causes of disease, carefully to adhere to the above-mentioned [Newtonian-Baconian] method of philosophizing. . . . Accustomed from the natural turn of his mind, as well as from the course of his education, to rest his belief on evidence alone, and to receive as true nothing but thus supported, he could not assent to theories built on round assertion, without the shadow of evidence to support them.

Immediately after these lofty words, Cooke proceeded to develop a single-entity theory of disease, based upon the notion that all human ailments were caused by an accummulation of blood in the veins of the liver and other abdominal viscera. The prescribed treatment, in every case, consisted of administering cathartics or bleeding, always undertaken as a simple mechanical operation, for he viewed the body as essentially a framework for a hydraulic apparatus, which operated on mechanical, never on chemical, principles. The fact that he could follow his appeal to Bacon by the development of such a full-blown a priori system of medicine, without "the shadow of evidence" to support it, should not be taken as indicating his insincerity; it merely provides a good example of the ritualistic nature of the affirmation. At any rate, Cooke remained true to his system to the end, for it is recorded that during his final illness he bled himself copiously and purged himself thoroughly with calomel.[4]

Even though an appeal to the name of Bacon was a mark of

scientific orthodoxy, it is difficult, in reading the works of nineteenth-century scientists, to determine exactly what they meant by "the method of Newton and Bacon." And the example of men like John Esten Cooke, who were convinced that they were Baconians, makes it especially difficult. They certainly did not have in mind the elaborate rationalistic pattern actually outlined in the *Novum Organum;* nor were they thinking of the mathematical reasoning of Isaac Newton. What they meant by "Baconianism" or, alternatively, "Newtonian philosophy" was a great deal simpler than either Lord Verulam or Sir Isaac would have recognized as sound philosophy. They were struck by Bacon's eloquent appeal for the study of facts as opposed to idle speculation, and, despite the apparent contradictions in practice, this is what Baconian philosophy meant to them.

The vagueness that was allowed to remain in the concept was an essential part of nineteenth-century Baconianism. At least three different meanings, none of which had to agree with each other, are hidden beneath this formulation, and such vagueness explains both the power of the term and a great deal of the confusion in the scientific thought of the century. First, and most evidently, "Baconianism" meant "empiricism," in the sense that all science must somehow rest on observation—that it must begin with individual facts and pass gradually to broader and broader generalizations. Newton or Thomas Reid or even William Whewell could with justice be called "Baconians" in this first, although trivial, sense. Secondly, it meant "anti-theoretical," in the sense of avoiding "hypotheses" and not going beyond what could be directly observed. Most nineteenth-century scientists considered themselves Baconians in this sense, although from the perspective of a century it is often difficult to follow their justifications. Third, and most radically, it was the identification of all science with taxonomy. Bacon's nineteenth-century followers in America usually had in mind the second or third meaning when they invoked the name of Bacon; the "nommer, classer, et décrire" of the French anatomist Georges Cuvier.[5] And their consistent failure to define the concept made it possible, for example, for them to call Newton a "Baconian" as he was in one sense of the word, and then allege that he was "Baconian" in all senses of the word.

There was little need for either definition or elaboration when the rules were so simple. "The object of the inquirer into physical science is generally regarded as twofold," as John C. Warren, prominent Boston surgeon and medical journalist, put it: ". . . first, the careful observation of facts—and secondly, the comparison of these facts so as to deduce from them the laws of the science in question." It was evident, he thought, that the first step ought to precede the second, and there was little more to say about the matter.[6] To go further would be to unnecessarily belabor the obvious, or perhaps to become lost in endless—and fruitless—philosophical problems, as the "sophists" of the previous century were widely held to have done.

If one does not read unwarranted subtleties into the use of the word "comparison," that is, if he understands it as simply implying *classification*, this completely mechanical—indeed, automatic—process described by Warren is a good summary statement of the third meaning of nineteenth-century Baconianism. The key word is "deduce," which is to be taken quite literally. One begins with a completely unprejudiced mind, observes a sufficient number of facts, and by simply comparing the facts with each other "deduces" the relevant law. And in this process one is to remain absolutely free of "hypothesis," a common term of disapprobation at the time. In other words, nineteenth-century Baconianism, as most American scientists used the term, implied a kind of naive rationalistic empiricism—a belief that the method of pure empiricism consistently pursued would lead to a rational understanding of the universe. American Baconians agreed with Scottish philosophers in the assumption that the intuitions of the mind were direct, immediate perceptions of a real objective order. The testimony of the senses had to be admitted as true, and its validity depended upon no outside, additional evidence. Joining with Thomas Reid, founder of the Scottish school of philosophy, in denying the actuality of the abstract particular—in the belief that mind and nature are so joined that even the initial sense impression contains reference both to a permanent subject and to a permanent world of thought—American Baconians generally held that error could not possibly arise in the observation of facts. Error could only be introduced by false or too hasty inference from the never-to-be-doubted facts and could therefore be corrected by additional observations. The truthfulness of the testimony of the senses could not even be questioned, as one spokesman said, "without questioning the truthfulness of our constitution, nay, the veracity of God himself—without questioning everything, through whatever channel derived." [7] Both the Scottish philosophy and the "Divine Veracity" guaranteed that a "sufficient number" of observations, in themselves, would yield a law of nature.

But the scientists themselves never pursued the subject any further than this bare outline, this brief declaration of faith. And it was nearing the mid-century before anyone felt called upon to make a serious defense of the Baconian philosophy. Before this time one could quite justly say, with Dugald Stewart, that the "merits of Bacon, as the father of Experimental Philosophy, are so universally acknowledged, it is superfluous to touch upon them here." [8] By 1847 one would not feel the superfluity quite so much, for as one American writer observed, it had become quite fashionable of late to deprecate "the lofty pretensions of the Baconian philosophy." [9]

During the decade of the 1830s—at the very time a scientific orthodoxy was being established—attacks on Baconianism's "lofty pretensions" had been launched from three different directions. Sir David Brewster had argued that Newton was the true father of modern experimental science, and that the method of Bacon was "never tried by any philosopher but Bacon himself," and the example he had given of its application "will remain to future ages as a memorable instance of the absurdity of attempting to fetter discovery by any artificial rules." [10] Brewster denied what most earlier students had affirmed, that Bacon and Newton had used the same method; more seriously, he even denied that the method of Bacon was anything like the proper method of science.

The second attack was the direct opposite of Brewster's. Macaulay had argued that Bacon's analysis of the inductive method was "not a very useful performance," because "it is an analysis of that which we are doing from morning to night, and which we do even in our dreams . . . it has been practiced ever since the beginning of the world by every human being." Even Macaulay's limited praise of Bacon could be taken as an insult, for according to him Bacon's only claim to a place in history was the fact that he originated the utilitarian movement.[11]

These charges were both serious, and since they were from respectable authorities they were disturbing. According to one, Bacon's philosophy was not only virtually untried but absurd and impracticable; and according to the other it was surely a correct description of what happened in reasoning, but it was trivial. For the first time the facile assimilation of the thought of Bacon, Newton, Locke, and the Scottish philosophers was being challenged, and such a challenge threatened the old assumption that nature's rules were clear, and the correct procedures obvious. As serious as these charges were, however, they were not so devastating as a third, coming from an entirely different quarter. Prominent Roman Catholics, both theologians and laymen, and also High Church Episcopalians, were beginning to look askance at the Baconian philosophy of science. Bacon was an atheist, charged Joseph De Maistre, and what was more, his philosophy led inevitably to atheism. De Maistre, in his two-volume examination of Bacon's philosophy, argued that among all the blasphemies which his century could offer against good sense, morality, or the dignity of man, there was not a single one that could not be found in the works of Bacon.[12] De Maistre's work was a long and scholarly complaint concerning the dominance of science over religion, most unsettling for its obvious implication that the two fields were not, as the Baconian mind would have it, complementary. Continuing the attack on science in general, an American Catholic writer exclaimed in evident irritation, "the fact is as common as it is deplorable, that a man may have penetrated deeply into the abstruse questions of mathematics, and not have devoted a serious thought to matters of religion." [13] Another Catholic commentator regretted that physical science, instead of coming to the aid of theology, was attempting to confound it.[14]

Thus, in the course of a single decade, there arose three essentially different, clear attacks on nineteenth-century Baconianism. There were at the same time other threats to the philosophy that were not quite so explicit. German transcendentalism, for example, was recognized by some as being opposed to belief in the primacy of experience, but this was neither so clear nor so influential a threat.[15] At any rate, for the first time since its widespread acceptance in the seventeenth century, the Baconian philosophy was being seriously challenged by extremely vocal individuals who could not simply be dismissed. Coming on both scientific and moral-religious grounds, the criticism threatened the very foundation of the dominant philosophy of science, and it was a threat to the standing of those who claimed to be followers of that philosophy. It was out of this context that there arose the first systematic American exposition of the Baconian philosophy which had been assumed without either question or elaboration by the preceding generation of scientists. As long as Baconianism could be thought of as partaking of the axiomatic there had been no need for either defense or explicit enunciation, but in the 1840s, when it was apparent to many observers that the day of Lord Verulam was passing, "defenses," "restatements," and "answers to objections" began to pour in from all sides. Between 1840 and about 1870 an extremely large literature of Baconian apologetics was developed.[16]

The foremost apostle of Bacon in America during the decade of the 1840s—one of the earliest and, in his own time, most influential contributors to the rear-guard literature of defense was not a scientist but a Maryland attorney, Samuel Tyler, whose ideal qualifications recommend him as an authentic spokesman for his generation. Throughout his lifetime he was highly acclaimed by his contemporaries as the greatest philosopher America had yet produced, and it was suggested more than once that his name would go down in history beside those of Aristotle, Bacon, Locke, and Thomas Reid—a curious assemblage perhaps, but nevertheless typical mid-nineteenth-century nominations for a philosophical hall of fame. In 1847 the *Princeton Review* pronounced it an "intellectual feast" to peruse Tyler's theoretical writings, and declared that the true principles of thought had not been either so accurately investigated or so lucidly stated by any author of the age as by Tyler in his *Discourse on the Baconian Philosophy*. In 1859 the editor of the *Southern Presbyterian Review* thought Tyler was "the first philosopher in America," and in the same year Francis Lieber assigned him a place among the most profound systematic thinkers. Sir William Hamilton was so convinced of Tyler's merit that he urged him to give up his law practice in order to devote all his time to philosophy, and in 1877 Tyler produced a final revised edition of his *Discourse* at the request of Joseph Henry, who thought him the ideal man to provide a text on scientific method.

A comment by the editor of the *Princeton Review* reveals very well the reason for Tyler's popularity, and at the same time it helps explain why he has been so completely forgotten by history. After bestowing his high praise on Tyler, the editor candidly admitted: "This, perhaps, will be considered as saying no more than that his views and reasonings are more accordant with those which we entertain than any other author with whom we are acquainted." The editor further suggested that Tyler would be doing a great service for humanity if he were to supply an elementary psychology text, which was sorely needed in order to overcome the "fanciful flights of Coleridge, or the vague and misty transcendentalism of the German school." [17] Tyler, in short, was not at all out of step with the dominant thought patterns of his period. History could not treat kindly a man so thoroughly conventional, but the historian who wishes to investigate the commonly held ideas of an intellectual class could find no better example. More than anyone else of his time, he managed to sum up the thought about science that dominated America in the first half of the century.

Samuel Tyler read his first book of philosophy, Reid's *Inquiry on the Principles of Common Sense,* while a student of law. With the grounding in common-sense philosophy provided by the Scottish writer's volume, Tyler was ready to publish a criticism of the next philosophical work he read—Cousin's *Introduction to the History of Philosophy.* During this same period he published a bitter review of Balfour's work on universal salvation in the *Princeton Review* for 1836, and an essay on natural theology in the *New York Review* the following year.

The ingredients of Tyler's thought were present and well-formed several years before he began his series of articles on the Baconian philosophy. The main elements were apparent in his earlier critical work—a passionate attachment to the Scottish common-sense philosophy of Reid and Stewart, a corresponding hatred of things "metaphysical" [18] (especially if they were either French or German) as he showed in his criticism of the Frenchman Cousin, and the deep commitment to conventional religion, especially of the Evangelical variety, which was evident in his review of the universalist Balfour and also in his work on natural theology. These were the ingredients that he would later shape into a philosophy of science that would make him the belated spokesman for that rapidly disappearing generation of Baconians.

A man of the early nineteenth century with Tyler's particular pattern of beliefs and prejudices could hardly be anything but a Baconian—a proposition that Tyler devoted the major portion of his life to demonstrating. It is not known exactly when he first read the works of Bacon, but it is clear that it was after he had thoroughly mastered the Scottish philosophy, for he, like most other Americans of his time, saw Bacon through the eyes of Thomas Reid and Dugald Stewart. At any rate he seized upon the philosophy of the *Novum Organum* with a zeal that was only matched by his industry in writing articles about it. Although there exists no published list of Tyler's writings, one sees his work everywhere in the reviews of the 1840s—in Breckinridge's *Baltimore Literary and Religious Magazine,* in the *American Journal of Science and Arts, the American Quarterly Review,* and even in the Baltimore newspapers, but above all in the pages of the *Biblical Repertory and Princeton Review.* This analysis is based primarily upon a series of articles published in the last-named journal beginning in July, 1840.

The first article, nominally a review of Basil Montague's edition of Bacon's works, had as its purpose "to exhibit the nature of the logic taught by Aristotle, in his *Organon,* and the nature of the Method of Investigation taught by Bacon in his *Novum Organum.*" [19] First of all, he assured his readers that there was no necessary conflict between the Aristotelian syllogism and the Baconian induction; they were, in fact, complementary. Francis Bacon had never intended to cast aspersions on deductive logic per se, for certainly he realized that Aristotle's logic was simply a description of the way the mind worked when it was functioning properly. A syllogism had to be present in all instances of correct reasoning. Induction, on the other hand, was not a process of reasoning at all, but a method of investigation and of collecting facts and phenomena; an activity that had to precede any meaningful reasoning. It was the neglect of this first step, that had been the real failing of all philosophy before the time of Bacon. The Greeks and their medieval followers began with general principles which had not been established by appeal to nature; pointing out their error and showing the world the way to move beyond the Greeks was the monumental contribution of Lord Verulam. In a word, before Bacon's time philosophy had not necessarily been wrong; it had simply been incomplete.[20] Bacon's message to the world could be summed up in the proposition that "the whole of philosophy is founded upon experience, and is nothing more than a classification of the facts and phenomena presented in nature."

In making explicit this identification of scientific method with taxonomy, Tyler was voicing a conviction generally accepted by his contemporaries. Although isolated examples of dissent could be found,[21] majority opinion was certainly on his side. From this viewpoint, the work of the scientist was merely that of collecting particulars and grouping them into classifications of the lowest degree of comprehension; and, after further observation and collection of facts, grouping them into classifications of higher degree until finally he would arrive at classifications of the highest degree, presumably ending at the one class of all classes.[22] The pervasiveness of this identification is suggested by the high praise generally accorded by the American reviews to taxonomists like De Candolle, and the corresponding neglect of the morphologists, who were currently doing important work especially in Germany. George B. Emerson, President of the Boston Society of Natural History and therefore official head of the Boston scientific community, said of De Candolle that "few works give us better executed examples of the Baconian method," which he said was that of forming general conclusions from the careful observation of particulars and then going back to reexamine the particulars. Emerson described De Candolle's procedure as follows:

He first minutely examined all the species of a genus, and thence drew his generic characters. From a similar full examination of all the genera of an order, he drew the ordinal characters. This done he returned to the genera and species, and rejected from the generic what had been sufficiently expressed in the ordinal, and from the specific, what had been distinctly stated in the generic characters.[23]

De Candolle was, indeed, an important figure in the history of botany, but what is of special interest is that Emerson explicitly identified his work as the model of all scientific investigation, thus agreeing with Tyler on the nature of science. Presumably any work that could not be placed in the precise framework of botanical taxonomy could be rejected out of hand as "unscientific."

After his brief definition of induction, Tyler turned in the same article to a consideration of the nature of inductive evidence. Since he thought that science was really classification, the evidence had to be concerned with the way phenomena were compared with each other and placed into categories. Comparisons could be made only on the basis of identity or analogy, the latter being any real resemblance less than identity. Notwithstanding the objections of the eighteenth-century philosophers, Tyler held that analogy was not only admissible but actually the most valuable kind of evidence. "What we mean by inductive evidence," he insisted,

is evidence found in the constitution of nature—real evidence, as opposed to mere hypotheses. . . . clothed by nature with the power of producing conviction in our minds, when it is fully apprehended, even in spite of ourselves. As to the first point, that analogy has a real foundation in nature, no one can object; for we can trace it everywhere. And as to the other point, whether it is clothed by nature with power to produce conviction in our minds solid enough to be the foundation of sound inductive inferences, we think there will be little objection, after diligent inquiry into the matter.[24]

Tyler went on to give several instances of analogical thinking— Wells' theory of the dew was cited as the model of all inductive investigation, and Newton's generalization of universal gravitation was also claimed for analogical thinking.

Tyler's two bases for comparison, without further clarification, really amounted to saying very little—things are either entirely like each other, in which case a relation of identity exists, or they are alike in some respects, in which case the relation is said to be one of analogy. The indeterminate character of the second basis will receive special attention in a later context, but for the present purpose it is only important to note that a particular psychology is clearly implied, indeed demanded, by Tyler's discussion of inductive evidence. The crucial tests for the admission of analogical evidence were: (1) that we can trace analogy everywhere, and (2) that it is clothed by nature with the power of producing conviction in our minds. Both of these tests have to do, not with an external nature, but with the minds of men—*we* can trace it everywhere, and it produces conviction in *our* minds. In short, the most important function of these tests for evidence was to remove the empirical problem from the realm of physical nature and make it simply a problem of minds—of correct belief. In their anxiety to avoid the skeptical conclusions of Hume, the Scottish philosophers had so blurred the line between metaphysics and epistemology that Tyler, reasoning on their principles, was able to argue that the basis of a philosophy of science could be found in a psychology of perception. What is more, Tyler himself knew that a certain kind of psychology was demanded if one were to avoid the obvious subjectivist implications of his evidential tests, and three years later he was ready to make the connection explicit. He found his psychology ready made in that of John Locke "as interpreted and modified" by Reid and Stewart.

The vehicle for his exposition was a review of Locke's *Essay Concerning Human Understanding and* Dugald Stewart's edition of Reid's works, published thirty years previously. Again Tyler was direct about his intentions. He referred to his earlier sketch of the Baconian philosophy and said that he now proposed to give an outline of the psychology or theory of mind assumed in that method; "and which the influence of that method upon English philosophy has caused to be developed by Locke and Reid." [25]

Tyler began his exposition by postulating that the Baconian method depended upon the psychological doctrine that all our knowledge is founded in experience and is acquired through sensation and consciousness. Locke had been falsely interpreted as having subscribed to the Gassendist doctrine *"Nihil es in intellectu quod non fuerit in sensu,"* but as Stewart had shown, Locke had only opposed innate ideas in the Cartesian sense—that they were coeval in existence with the mind to which they belonged, and therefore they illuminated the understanding before the external senses began to operate. Locke had never, even in his zeal to refute Descartes, excluded innate operations of the mind, and therefore he had to admit the introspective experience of consciousness as a source of ideas. Once this misunderstanding had been cleared up, Locke's psychology could be seen as the direct correlative of the Baconian method.

The Creator of all things, Tyler said, had established an order, an antecedence and sequence, of both matter and mind in the universe and the sole object of philosophy was to discover this order. The belief that such an order existed in both universes was a fundamental postulate of all philosophies, he continued, for philosophy would be futile without the concept of order. But serious disagreements arose concerning the manner in which the two universes interacted in perception. The "idealists," of whom he cited Hume as the most notorious, had by their corrosive skepticism undermined the idea that man perceived nature directly and had seemed to demonstrate that the mind perceived images of things and not things themselves; the mind was an ordering mechanism which gave no guarantee of faithful representation. This raised the question of whether one could legitimately make any statement about the world, as distinguished from statements about a sequence of images in the mind which had no necessary connection with things "out there" in the world. According to Tyler, the doctrines of the "idealists" would work a doubly pernicious influence, for they had clearly corrupted both philosophy and revelation during the previous century. One found that the same kind of errors appeared in the theology of nations which had adopted the theory of innate ideas, as in their philosophy. In the one case, they considered philosophy the development of human reason; and in the other they submitted the truths of Divine revelation to the judgment of "innate moral principles." [26]

Philosophy had continued in its state of corruption until finally a "champion of the truth appeared," in the person of Thomas Reid of Edinburgh. Attacking the central conception of the idealists, Reid showed that their argument that the mind perceived images of things rather than the things themselves, when applied to any other sense than sight, was "perfectly absurd," for one could not even conceive of what an image would be as applied to, say, the sense of touch. Given the assumption, which Scottish philosophers demanded, that there was a direct connection between mind and nature, it followed that that which could not be conceived could not exist.[27]

After demolishing the "idealists" Reid's next contribution was to establish three laws of the mind, which Tyler summarized as follows:

such is the constitution of human nature, that man cannot but believe in the reality of whatever is clearly attested by the senses . . . . such is the constitution of human nature, that man cannot but believe in the reality of whatever is clearly attested by consciousness . . . . man is so constituted that he cannot but believe in whatever he distinctly remembers.

Reid's "laws of the mind," in short, were designed to overcome all the difficulties of philosophy by dismissing them as irrelevant; for the philosopher could find anything he needed as a constituent of his system by simply assigning it to the "constitution of human nature." But Reid's greatest contribution, according to Tyler, was yet to come. He had still to overcome the skepticism of Hume in regard to causes. The procedure in this case was exactly the same:

Reid therefore showed by a most rigid analysis of mental phenomena, that man is so constituted that he cannot but believe that like causes will produce like effects; and that the future will be like the past: and thus discovered the great fundamental law of belief which governs the mental determination in the inductive process; and thereby evolved another point of affiliation and doctrinal identity between the English psychology and the Baconian method of investigation.[28]

In terms of the thought patterns of the period, Reid's argument was a good one. Resting the critical part of his argument on the test of "conceivability," and the constructive part on innate characteristics of the constitution of man, Reid was taking advantage of two points generally agreed upon. Even by those who rejected the major portion of Baconian philosophy, it was still believed that if one could not even "conceive of" the necessary consequences of a doctrine, the doctrine could be summarily rejected. Obviously, one cannot "conceive of" an image of anything other than the sense of sight. By the same token, Reid's use of innate characteristics of the human constitution could be made appealing in terms of the empirical predilection of the generation, for those were matters that every man could put to the test himself. Tyler was not, then, being peculiarly naive when he reported that he was "completely satisfied" with Reid's argument. His only hope was that the "transcendental speculations of the German and French philosophy," which, he noted with a measure of apprehension, were becoming more and more popular, did not cut the English philosophy loose from its "strong anchor of common sense . . . to be dashed and tossed, by every wind of speculation, upon the boundless oceans of skepticism." But yet, perhaps in preparation for a later article, he capped the argument by showing that the doctrine that all our ideas come from experience accorded with scripture. It was distinctly taught in the book of Genesis, he observed, that the first man received the truth by immediate revelation from God, and that by a "supernatural tuition" he was instructed in knowledge which he could not have acquired by his unaided intellect. This demonstrated, for example, that even the idea of God was ultimately based upon experience.[29]

It becomes quite obvious in considering Tyler's arguments against the "idealists" that the only difference between them was that Tyler had derived from the Scottish psychology of perception an absolute guarantee that the mind would give *true* representations of nature; this belief would not have been tenable without a commitment to the proposition that God had erected a parallelism between mind and nature. It was God who had provided the necessity behind those laws of belief, and since God could not be responsible for a lie, that which conformed to man's mental constitution had to conform to the physical world. In a word, at the bottom of Tyler's science was a theology. In the third article of the series, Tyler's theological underpinnings became explicit.

In this review, which was occasioned by the American edition of Montague's *Works of Lord Bacon* (1842), Tyler developed no new ideas concerning the Baconian philosophy; he was concerned with the influence of Baconianism in the modern world, and he wished to vindicate the philosophy against some of the charges which had been leveled against it within the past decade. Now that he had developed the basis of the Baconian philosophy and had shown how it had to result in truth, he would attack the enemy directly.

Tyler could sum up the influence of Baconianism quite readily; it had made England, the home and still the most assiduous cultivator of this philosophy, the leading nation of modern times. It was no accident, he thought, that England was not only the "great progressive and regenerative nation of modern times," but it was also "eminently conservative," possessing in an unusually happy combination the elements of both progress and stability.[30] In every department of physical science England had made the leading discoveries. Modern physiology had begun with Harvey, modern medicine with Sydenham, and comparative anatomy with John Hunter. Beginning with the work of the incomparable Newton, the most important discoveries in physics had been made by Englishmen, and the same was true of chemistry. The work of all these men, of course, provided examples of pure Baconianism. Aside from its purely scientific benefits, the legacy of Lord Verulam had done more to develop the wealth of England than "all the legislation of all the statesmen who have adorned her history."[31]

But in spite of all the remarkable advances that could be attributed to the Baconian philosophy, Tyler said, certain misunderstandings had arisen concerning it. The doctrines of Hobbes, Hume, and Hartley in England, and those of "the host of infidels and atheists in France," had again and again been proclaimed as the legitimate deductions from Bacon's philosophy. In arguing that the deductions were not legitimate ones from a correctly understood Baconianism, there were three specific charges that most concerned him. In the first place, he answered them all in general by noting that the real excesses which people pointed out were not the errors of Bacon but the result of his misguided successors seizing upon some isolated part of his philosophy and carrying it out to the wildest extremes without reference to the whole of the philosophy. Such were, therefore, not a part of the Baconian philosophy, but merely the "exuviae" thrown off from it as it had passed through the process of development. Cicero, he noted, had remarked that the very same thing happened to the movement started by Socrates; it was, in fact, what commonly happened to the philosophies of all great men.[32]

Turning to the three charges in particular, he began with the assertion that Bacon's philosophy was sensual and utilitarian and that it necessarily led to a selfish moral philosophy. This argument was easy to dispose of, Tyler said, for it was clearly based upon a misconception. Those who made this charge rested their case on the alleged fact that Baconianism admitted but one source of ideas—sensation—and they correctly argued that within the sphere of sensation there was no idea of right and wrong, but only of pleasure and pain. But the Baconian philosophy, as Tyler had shown before, admitted both sensation and consciousness—it was, in fact, dependent upon the postulate of consciousness. Since it was on the latter that the foundations of morality were laid, one could not blame the Baconians for the excesses of pure sensationalism. According to the precepts of inductive philosophy, one must examine all the facts of man's moral constitution and establish the fundamental truths of moral philosophy by psychological observation. Rejecting all innate moral principles or notions, it left man perfectly free to examine "all the facts of his moral constitution" and to establish whatever system of morals a sound induction warranted. There was no doubt what kind of moral system such an objective examination would produce, Tyler assured his readers, because a careful study of the heart of man would show the presence of certain instinctive affections, such as love, hope, fear, anger, pity—all disinterested in their nature—and in the mind, one would find the power to distinguish moral good and evil. Since it was upon these attributes of the spiritual nature of man that both Baconian philosophy and Christianity founded morality, the argument, with which Tyler agreed, that the sensations knew only pleasure and pain was completely irrelevant.[33]

The second charge could likewise be readily dismissed, for it was based upon the same misconception. It had been said that the Baconian philosophy had no ideal, and that it was utterly destitute of the sense of beauty. Tyler merely referred here to his previous argument that Baconianism admitted consciousness, for in admitting consciousness it evidently had to recognize all the sensibilities on which the science of aesthetics was based. But on this point Tyler thought that there was really no need to deal in philosophical analysis or controversy; one had only to appeal to history to show that the philosophy of Bacon was not only consistent with the arts of beauty but highly conducive to their advancement. "Where then," he asked rhetorically to clinch the argument, "is there a nobler literature, than that which has been cultivated in the same soil and by the same people, with the Baconian philosophy?" This coincidence, Tyler urged, should be enough to confute even the most obtuse skeptic.

As for the third charge, which Tyler recognized to be a much graver one, he cited De Maistre and Schlegel to the effect that Baconianism conduced to materialism and atheism. He argued against this charge on much the same grounds as he had the others—that consciousness informed us of the existence of an immaterial soul, and that no other people on earth had cultivated natural theology so much as the English. After all, had it not been an Englishman and a perfect exemplar of Baconianism, Isaac Newton, who had brought men back to thinking in terms of final causes after the French infidel Descartes had endeavored to banish them? The works of Bishop Butler, Paley, and the more recent Bridgewater Treatises indicated the continuing interest of the English in natural theology. With all this evidence before him, if the skeptic would merely survy the works of the English dispassionately, he would have to conclude that philosophy, revelation, natural theology, and physical science "are united in perfect harmony, proclaiming with one voice that there is a God."

After showing that all the arguments against Baconian philosophy were unfounded and based upon misconception, it remained only for Tyler to sum up its real character in a single sentence: "It embraces every thing that is sublime in speculation, useful in practice, lofty in morality, beautiful in art, and reverential in religion." [34]

Tyler had alluded to religion in every one of the articles and in all the others he wrote on the subject, but his opportunity for a full-length treatment of the religious aspects of Baconianism finally came in a review of the Reverend John Stone's *The Mysteries Opened* (New York, 1844). The book under review, he noted, had been written by the Reverend Mr. Stone in order to put down the "superstitious errors of monkish theology," which he noted had been refuted many times in the past, but were still reappearing in the field of Protestant theological controversy, even in the midst of all the enlightenment of his own age. Tyler thought that Stone's book was an irrefutable answer to the Oxford reformers, who were the main proponents of "monkish theology" in the Protestant world, but he hardly felt that there was any need to outline its argument for the Presbyterian readers of the *Princeton Review*. Instead of repeating these arguments for the already convinced, he said he would use the occasion to examine what he conceived to be the groundwork of all theological controversy; the connection between reason and revelation. Because of Tyler's own commitment to the proposition that there were no innate ideas, he argued that if one used the word "reason" meaningfully it could only signify "philosophy." Any other use of the word would imply "rationalism," which he understood to mean "working from accepted a priori principles, a procedure he considered so absurd that it could not possibly have anything to do with religion. Therefore, the question that he would examine resolved itself into "what is the connection between philosophy and revelation?"; that is to say, "What assistance does the light of nature afford us in examining the truths of revelation?" [35]

Tyler began the analysis by stating an important limitation to the utility of philosophy. The "light of nature" could not be used to prove scriptural doctrines, for this was contrary to the fundamental idea proclaimed in Christianity. If one presumed to test the truths of revelation by philosophy, it would destroy the whole idea of revelation as a direct, God-given truth. And science, of course, could not be inimical to revelation if correctly understood. All a priori systems, however, were in necessary conflict with the basic principles of Christianity, which were based solidly upon the concept of an inerrant Scriptural revelation. The a priori philosophers would submit these truths of revelation to the judgment of some mystically conceived "reason," and in doing this they would make the mind of man the judge, not merely of the *substance* but of the *validity* of the

Divine revelation. Such a procedure, he thought, was not only bad philosophy but alien to the spirit of Christianity. It was, in fact, no better than "atheism," or, at the very least, was playing into the hands of atheism. It implied an assumption either that all of the Bible was not Divinely inspired, in which case there could be no certainty concerning any part of it, or that God might lie, in which case both science and religion would be impossible.

But even though the basic truths of religion could not be submitted to the light of nature, this did not mean that philosophy was irrelevant to religion. Quite the contrary, Tyler argued:

there is a philosophy which is consistent both in its method of investigation, and its principles with Christianity—a philosophy, which, humbling itself before Christianity, acknowledges it to be a revelation of a knowledge that lies beyond and above its province. This is the Inductive or Baconian philosophy.[36]

Part of Tyler's argument was the same as the one Bacon had made—simply that theology is grounded only upon the word of God and not upon the light of nature. But this is by no means all that he meant. Although philosophy could not judge the word of God, it could still, he thought, settle theological controversies; that is, if pursued along sound Baconian lines. This was so simply because Christianity was a *finished* revelation, and all its doctrines were found in a Book which was readily available for anyone's inspection. The reason that there had been so many perversions of scripture and so many false doctrines taught in the name of Christianity was that the Bible was not a work of systematic theology; its truths were not arranged in any particular order, and they were seldom entirely clear in any single part of the text without reference to the rest of it. The great principles upon which true religion was based were scattered about in both Testaments as God in his own time had chosen to reveal them. Given such a situation, it was extremely important to consider every reference to a subject, and it was equally important to consider each item of doctrine in relation to the whole.

A priori philosophers, however, had seldom been careful in this respect. Instead, they had been in the habit of seizing upon some passage that struck their fancy, and in isolation from the remainder of scripture giving it some interpretation that squared with their "reason" or with "nature"; and then they would try to make the rest

of the revelation conform to it. Tyler held that this kind of interpretation was a travesty upon the revelation, and it could be completely avoided by a judicious use of the Baconian philosophy—the
only approach which could give the properly reverential mode of
interpretation. One should first make a sound use of the rules of gram-
mar and logic, and in the approved Baconian fashion collect all the
passages on the same subject matter. Then, from the induction of the
whole, one could deduce the meaning of each, rather than infer the
meaning of all the rest from the interpretation of one passage which
the individual might fancy to be the leading one. There was, of
course, one possible exception to this rule; where the meaning was
so plain, "like what Bacon calls 'glaring instances' in nature," one
might use that passage as the key to the meaning of less obvious
passages on the same subject. "In a word," he said, "we must make
scripture the infallible rule of interpreting scripture; just as we make
nature the infallible rule of interpreting nature." [37]

Thus it seemed to Tyler that Christianity practically presupposed
and at the same time was presupposed by Baconianism. Biblical
exegesis was only one more example of a problem in classification;
and since this was so, it could be handled exactly like any other
scientific problem.[38] At this point, Tyler had elaborated a philosophy of science which, although still following Baconian lines, was
more suitable than the original to sustain the other beliefs nineteenth-century thinkers found important; and in turn, it was tem-
pered by those other beliefs. Where Bacon had intended classification to be the beginning of science, the naive realism of Scottish
philosophy could be used to make it appear that this was the whole
of science. And where Bacon had wished to separate religion and
science, the injection of Scottish philosophy could make it appear
that the two were inseparable.

Bacon's *Magna Instauratio* was never completed, and what he
himself thought science ultimately was must forever remain a matter
for scholarly debate. It is not necessary to make a judgment on this
issue, however, in order to observe that the "Baconianism" of
Tyler—and the same could be said for most American thinkers of
that period—appeared to owe fully as much to the Scottish common-
sense realism and to evangelical religion as it did to Francis Bacon.
This is not to suggest that Tyler was not well acquainted with the
works of Bacon, or even that his philosophy of science should not be

termed Baconian. Indeed, one could certainly argue that Tyler's philosophy was derived from Bacon; but one could with equal justice trace it directly from Scottish common-sense realism and evangelical religion. All I wish to do at this point is to make it clear that some of the basic assumptions of that period—if Tyler can be taken as a model—were inextricably tied up with maintaining belief in the philosophy of science called "Baconian." In arguing that Scottish philosophy provided the appropriate psychology for Baconianism, Tyler was making a case for both Scottish philosophy and the Baconian method, and his argument that Protestant Christianity was Baconian served the same double function. The same kind of errors appeared in the theology of nations that had adopted the theory of innate ideas, as in their philosophy, he warned. The legitimate fruits of German idealism in science, he had earlier told the pious readers of the *Baltimore Literary and Religious Magazine*, were "neology, transcendentalism, pantheism, and all sorts of mystical nonsense." Germany, he had then generalized, "is the hotbed of all imaginable strange doctrines." And this, of course, was a direct result of the a priorism of German philosophers.[39] The implication could not possibly have been more clear. Pious Protestants must, for the sake of their religion, at all costs maintain the Baconian Philosophy and help stamp out all rivals. That Baconianism was a clearly Protestant philosophy of science is well illustrated by De Maistre's attacks upon it and the rejection of it by American Catholic writers; the fact that leading interpreters of Bacon, like Samuel Tyler, were ardently anti-Catholic provides further evidence.[40]

Tyler himself drew more conclusions from his Baconianism than some of his Protestant supporters would have been willing to allow, for he was equally sure that it was not only a Protestant philosophy of science but an evangelical one—although it is not at all clear whether he was evangelical because it was the Baconian thing to be or Baconian because it was evangelical. At any rate, he closed the article on the connection between philosophy and revelation by observing that his analysis showed that the evangelical form of religion was the most Christian because it was the most Baconian.[41]

Tyler's suggestion that the inductive method could be applied to the settlement of religious disputes was more than the ardent High Church editor of the *True Catholic*, another Maryland lawyer, was able to read without protest. Hugh D. Evans, in a critical notice of Tyler's *Discourse on the Baconian Philosophy*, admitted that he had not had time for more than a hasty glance at the book, but he had seen quite enough to be convinced that it was a bad book. The disturbing thing he had noticed was that Tyler spoke of the Baconian philosophy as the philosophy of Protestantism. This heresy only convinced Evans that Tyler did not understand Protestantism as distinguished from rationalism. Although Tyler had campaigned vigorously against rationalism, which he understood as being the use of a priori reasoning, Evans, who understood rationalism as being simply "the attempt to reduce all things within the compass of the human intellect," was able to accuse Tyler of that sin. Evans had no quarrel with Bacon himself, for he thought it had been "the sophist Locke" who opened the floodgates of infidelity by applying the principles of Bacon to areas where Evans' religion told him they were clearly inapplicable. The Lockian psychology, which Tyler had connected to Baconianism, Evans thought to be pernicious doctrine, for it was simply not true that all our ideas were derived from experience. It was an especially fantastic claim when applied to religious matters. The proper method of considering religious questions, Evans argued, was to adopt the revealed facts as first principles from which to reason as axioms, not to set out to collect from an induction of particulars first principles for oneself.[42]

The difference between Evans and Tyler reveals very well the reasons why Catholics and High Church Episcopalians could not accept the usual interpretation of Baconian philosophy. It was quite acceptable, however, to most Protestants because of the method of Biblical interpretation and the way of settling theological controversies Tyler thought it implied. Tyler had said that the answers were there, in the closed revelation contained in the Bible, and any reasonable man could find the answer for himself by applying the Baconian method of collection and classification. This fit quite neatly with both the romantic-democratic assumptions of the age and, of course, with Evangelial Protestantism. "The Protestant Principle," as the *New Englander* observed in the same year that Tyler's book was published, was the right of private interpretation of the Bible as the only authority.[43] This principle had historically been denied by Catholicism, and at the time Tyler wrote it was being denied within Protestantism by the Oxford Reform movement, of which Evans was a prominent lay spokesman, and which had at least three eminent representatives among the Anglican clergy in America.[44]

But on the other hand, as T. E. Bond observed, "Methodism is *Religion without Philosophy*," and by *Philosophy* he made it clear that he meant the rationalistic a priorism that Tyler thought hostile both to the spirit of Christanity and to Baconianism.[45] This kind of "rationalism" (reasoning from accepted principles) would include the deistic rationalists, the Catholic writers, and the High Church writers, C. S. Henry of the *New York Review* and Hugh D. Evans of the *True Catholic,* even though the latter wrote two articles against what he termed "rationalism" (trying to work out one's first principles from inductive evidence) evidently as an attack on Tyler. The fact that both sides condemned something they called by the same name should not lead to the conclusion that their viewpoints were at all similar.

Since Bond's statement about Methodism could apply equally well to most Protestant denominations, and more especially to any evangelical religion, it is not surprising that the defenses of Baconianism were found in their journals; and that suspicion, deprecation, or hostility should be the dominant note in Catholic and High Church journals. It is no wonder that a Catholic writer could accuse Protestant scientists of passing off unfounded religious assertions in the company of "exact demonstrations" of scientific matters; [46] and it is no wonder that Baconianism found an enthusiastic home in nineteenth-century Protestant America, while it was received either with hostility or with mixed feelings in such European countries as France, Italy, or even England.[47] The fundamental Protestantism of America, and the ease with which the Baconian philosophy could be made to sustain that Protestantism, more than anything else explains the widespread attachment to Bacon. Since the philosophers of science were Protestants, and likely to be evangelicals, they were able to make it appear that God himself, who was a necessary postulate of their science, was both a Baconian and a Protestant. In early nineteenth-century America the two words often meant the same thing.

CHAPTER IV

⟨∿⟩

The Philosophy in Action

Although Tyler was the outstanding exponent of Baconianism in America, and may be said to have voiced a clear majority opinion on most matters connected with the subject, there is no record that he ever conducted an investigation along Baconian lines. His was, rather, the task of formulating and expressing the chief value-judgments of his period, and of showing how a particular method of scientific investigation was in keeping with those values. His Baconianism was, in a word, a rationalization of beliefs that were already widely held. But although rationalization was an important function of the philosophy, it was by no means its only one. On the contrary, as is common with many such rationalizations, there were men who lived by its principles, and there were others who were intensely uncomfortable because of their failure to do so.

We have already seen that botanical taxonomy was considered to be the model of Baconian methodology, and a number of other scientific disciplines, particularly those in a relatively primitive state, could be made to fit the framework. And as one would expect, it was this type of investigation in which American scientists of that period excelled. This is no doubt the reason why the natural historian rather than the physical scientist has come to be the popular image of the nineteenth-century American scientist. Their crude Baconianism had not equipped Americans to handle the more abstract problems, and their efforts in this direction were generally unfortunate.

If one would observe how this philosophy worked in practice, he must take leave of the philosopher of science and follow the investigations of the actual practitioners. Tyler was interested in providing a general framework—a rationale—for Baconianism; the practicing scientists were filling in the details as they collected and compared the data for their researches. Again, the Baconian philosopher in drawing his framework found no occasion to comment on methodological principles important for the Baconian investigator; for example, Tyler provided no analysis of the *experimentum crucis* or the *vera causa,* concepts of great practical importance. In order to understand how such concepts were applied, there could be no better figure to observe carefully than Elias Loomis, successor to Olmsted in the chair of natural philosophy at Yale.[1] The rationale for these principles, however, will have to be inferred, for Loomis, like most Americans of his time, thought that his principles were well-known and in need of no further justification.

Loomis' career is remarkably illustrative of Tyler's Baconianism in action. A widely learned man, Loomis was well-versed in both ancient and modern languages, in mathematics, astronomy, and natural philosophy. He was so interested in the collection and classification of facts that he even carried the method over into his hobbies, spending his later years preparing a definitive family genealogy—a favorite occupation, one might add, of elderly scientists during that century. To make the relationship to Tyler's Baconianism even more complete, after graduating from Yale in 1830 he spent two years attending the Andover Theological Seminary. Loomis is remembered in the history of science for two important incidents, both occurring early in his life, and both being concerned with the more primitive sciences. One had to do with terrestrial magnetism, the other with the action of storms. They bear looking into in some detail, for they show quite clearly the kind of questions that the Baconian method could answer. At the same time they reveal its fundamental limitations and provide certain insights into the failure of American science in the nineteenth century.

Men had noticed, at least since the time of Gilbert, that the magnetic needle was not quite so "true" as popular expression would have it. It had long been recognized that the magnetic compass varied from true North, and before a more satisfactory method (involving a chronometer) was devised scientists had attempted to use the variation as a way of determining longitude. Their efforts, of course, had ended in failure when a more sensitive needle showed that the longitudinal variation itself was subject to variation. Observers noted strange, apparently willful variations—jerky and irregular motions, motions with a daily period, motions with an annual period, and motions whose oscillations required centuries for completion. There had been a great deal of speculation about the cause of the variation, and scientists had not abandoned hope that the apparent wilfulness would turn out to be lawful—that is to say, could be predicted—even though the original practical need for such a determination no longer existed.[2]

Elias Loomis, becoming attracted to the problem of the unexplained variation, decided to make it the subject of his first major scientific investigation. He first undertook to investigate the daily motions of the needle—presumably regular variations that had first been observed by George Graham in 1722. His efforts along purely Baconian lines were indeed heroic. At the beginning of the second year of his tutorship at Yale College (1834), he set up by the north window of his room a heavy wooden block holding the variation compass belonging to the college. Since most people had assumed that the daily variation was connected in some way with changes in temperature, he hung a thermometer above the compass and recorded the temperature each time he observed the needle. For thirteen months Loomis faithfully watched the needle, recording its position at hourly intervals whenever possible, tracing its daily movement, trying to relate it to the movement of the thermometer, and painstakingly comparing the movements of nearly 400 days with each other. He had intended, he said, to make observations every hour of the day from five or seven o'clock in the morning until ten o'clock in the evening, but his tutorial duties had caused some unavoidable interruptions.

Even though his study was at this point inconclusive, Loomis was rewarded for his perseverance by having his observations published in the *American Journal of Science and Arts* in 1836.[3] Encouraged by the reception of his paper, he then resolved to press on to even greater things—to collect an even more impressive array of facts. He would do no less than draw upon all the observations of magnetic declination that hitherto had been made in the United States. Realizing he would need some help in this staggering task, he appealed successfully to the Connecticut Academy of Arts and Sciences. A call went out to watchers everywhere—army surgeons, surveyors, explorers, geologists, astronomers—and hundreds of observers rallied to the call. Mountains of observations poured in on the newly appointed Professor Loomis of Western Reserve University, and soon he was once again ready to publish. The fruit of this labor was impressive enough—he had conducted what was probably the most ambitious coordinated scientific investigation up to that time and he published the first map of the United States showing lines of equal declination of the magnetic needle.

By listing in a table, chronologically by areas of the country, observations he had collected, Loomis was able to draw a generalization:

that the westerly variation is at present increasing and the easterly diminishing in every part of the United States; that this change commenced between the years 1793 and 1819, probably not everywhere simultaneously; and that the present annual change of variation is about 2′ in the southern and western states, from 3′ to 4′ in the middle states, and from 5′ to 7′ in the New England States.[4]

Loomis' generalization fit the requirements of a Baconian induction perfectly, "flowing from the facts" as it did and going "beyond the facts" in no particular. To be perfectly precise, the generalization is a simple summary of the material in the table, and the map that he constructed was also a summary of the table. Tabular representation is of course convenient and initial caution is commendable, but the important point is that the table was clearly more than a convenience to Loomis—in his viewpoint an adequate table was the essence of science. And the modesty of the generalization was not an instance of initial caution; to Loomis a summary was the limit of legitimate generalization. His subsequent behavior strengthens the impression that he implicitly assumed that the only problem in science was that of constructing a better table from which better summaries could be drawn, for he continued his interest in terrestrial magnetism for many years, publishing revision after revision of his map, aiming only at more and more precision in filling out the table and in drawing the lines on the map.[5]

One should certainly not undervalue this kind of work. There is no doubt that something like what Loomis accomplished had to be done as a preliminary, for obviously it would have made no sense to make generalizations about magnetic declination before knowing, at least in a general way, what the declination was. But the important thing to note about Loomis is that it never occurred to him to advance beyond drawing those lines finer and finer. More observers simply meant more observations, better instruments more accurate observations; the only thing new in science was a new category in the table, resulting in a new line on the map, or a redrawing of an old one. What the lines might imply or how they could be explained was apparently of no interest to him. When he was satisfied that the lines had been drawn as accurately as he could make them, he abandoned the investigation and devoted his efforts to studying magnetic dip, in exactly the same manner, and with exactly the same end in mind. His observations on magnetic dip were carried out for several years and spread over thirteen states. And since various kinds of observational inaccuracies occurred with the dipping needles available at that time, Loomis took truly heroic steps to smooth out the errors; each determination he published was the mean of from 160 to over 4,000 readings.[6] The Baconian observations must be as precise as human ingenuity and poor instruments could make them.

There were some, even in Baconian America, who were not satisfied with this kind of descriptive knowledge of the phemonena —those who wanted to inquire into the meaning of Loomis' lines. One of these was John H. Lathrop, who published an article as Loomis was completing his work on the lines of declination. Lathrop had made the Baconian observation—as had Loomis—that the line of no declination gradually and regularly moved westward. Loomis, however, had called the westward movement a generalization and, having reached the end of measurement possibilities, had moved into another area of investigation. But for Lathrop the westward movement was the beginning, not the end, of science. It was simply a fact of experience and, like any other fact, in need of explanation. According to his calculations the phenomenon could be explained by reference to the igneous theory of the earth. The earth, according to the theory, had once been a molten mass and its present external appearance was therefore the result of a gradual cooling process. The external solid crust of the earth, according to Lathrop's authority on the subject, was even then only forty to fifty miles in thickness; the interior still being in a molten state. Applying the laws of mechanics to this theory, Lathrop then reasoned that this internal mass would be revolving in a westerly direction while the crust went through its normal motions. Given this hypothesis, one had only to suppose a complete revolution of the igneous mass every seven hundred years in order to account for the movement of the line of no declination. He noted that if the igneous theory of the earth were true the contraction caused by cooling would cause a gradual shortening of the day.[7]

Despite the speculative nature of Lathrop's article, it was nevertheless a real attempt to establish the connections between natural phenomena, which men like Loomis characteristically neglected to do. Lathrop's hypothesis was at least indirectly testable by mathematical demonstration, and he volunteered an empirical prediction which followed from the explanation. But no one paid any obvious attention to Lathrop's article; doubtless they noted the basic "fallacies" in his approach and decided to ignore the work. He had begun in an approved Baconian way by comparing data, although his comparisons over time, while not exactly prohibited by Baconianism, were quite unusual. As a matter of fact, when anyone attempted to use historical data he encountered embarrassing difficulties immediately, for he could never demonstrate that he had alleged a *vera causa* as the term was then understood. One of Newton's rules of philosophizing, which had become a part of the Baconianism of the period, required that nothing should be assumed as the cause of phenomena, unless it could be proved adequate to their production, and also shown to exist. This requirement, as Francis C. Gray had noted in a review more than twenty years earlier, was the downfall of all geological speculation, for "this requisition is not complied with by proving that there are such elements as fire and water; they must be shown by induction or analogy to exist in the place and the quantity, which the hypothesis demands." [8] In other words, mathematical demonstrations of plausibility and empirical predictions could not have mattered less; before he could be heard, Lathrop would have had to bring back a specimen of molten rock from the center of the earth and testify that he had empirically ascertained the extent of the igneous mass. And even this need not have satisfied the really determined Baconian, for there was still the stubborn fact of the solid crust. Perhaps it had simply always been this way. After all, the rate of cooling—if indeed it was a case of cooling—could not be measured.

It is entirely likely that considerations such as these explain why there is no indication that Loomis—or anyone else—noticed the article by Lathrop. Since it was published in the same journal in which most of Loomis' work appeared, he could not have missed the article—apparently it was simply beneath contempt and accordingly dismissed as just some more of the "nonsense" to which Silliman in his old age was increasingly opening his journal.[9]

Loomis' method achieved a notable, if somewhat limited, success in his first large-scale investigation, for he was asking the kind of question that Baconianism could very well answer. To the question: "How are the lines of equal magnetic declination distributed about the United States?" the method of collection and collation of data was admirably suited to find the answer; one had only to collect enough facts, arrange them in a tabular form, and the proper conclusion would immediately be evident—within the limits, that is, of measurement techniques and instruments. And in this kind of investigation, as Baconian methodology required, a larger number of observations, by smoothing out observer errors, could only make the answer more precise.

But the other important incident in Loomis' scientific life was a bit different. It had to do with the then young science of meteorology, in which Loomis had also displayed an interest since his graduation from Yale. In 1831, before Loomis entered upon his tutorship, there appeared the first of a series of papers by William C. Redfield, a New York naval engineer, upon the theory of storms.[10] In the last year of Loomis' study in Paris, there were published the first of a like series of papers on the same subject by Professor James P. Espy of Philadelphia.[11] These two series of papers represented two rival theories of meteorology, and they became the subject of a great deal of discussion at scientific meetings and in journals for a long period of time.

On his visit to Europe, before he went to Western Reserve, Loomis had obtained a set of meteorological instruments, and from that time on he regularly made complete sets of meteorological observations twice a day. By 1840 he was ready to attempt a decision between the two theories. Redfield's argument was that storms blow in circles counter-clockwise around a center that advanced in the direction of the prevalent winds. This argument had first been made in 1801 by Colonel James Capper of the East India Company, in a published work on the winds and monsoons of India. Redfield apparently reached his conclusions with no knowledge of Capper's previous work by studying storms in the Caribbean, and in spite of

Capper's prior suggestion of the theory, Sir David Brewster claimed for Redfield "the greater honor of having fully investigated the subject, and apparently established the theory on an impregnable basis." [12] In Europe the rotary theory was at the same time being advocated by Professor Dove of Berlin and by Lieutenant Colonel Reid of the Royal Engineers, who had made intensive studies of storms in the Barbadoes.

But despite Sir David Brewster's judgment, the "impregnable" establishment of the rotary theory was no more than apparent, for there still existed an influential and persuasive rival theory. M. Brande, working with European registers of a storm of December 24, 1821, concluded that the winds blew from all points of the compass toward a central space, where the barometer was, for the moment, at its lowest stand. This was the theory that Espy had adopted and further developed. Redfield's theory was generally adopted in England, having been endorsed by the Royal Society and the British Association; [13] while the French Academy endorsed Espy's theory. [14] Espy argued that observations showed a centripetal motion of the wind, toward a center if the region covered by the storm was round, and toward a central line if the storm region was longer in one direction than in another. Espy's conclusions were, in part, derived from his belief that in the center of the storm there was an upward motion of the air, and that the condensation of vapor into rain furnished the energy needed for the continuation of the storm. The main point of contention between the two could be reduced to this central question: Do the winds blow in circular whirls, or do they blow in toward the center?

The difference between the two theories did not rest upon different selections of evidence; on the contrary, Redfield and Espy drew their rival conclusions from the same reports of the same storms. The fact was that either theory could quite well explain the reported action of the storm. The evidence that they had to work with consisted, for the most part, of ships' logs, reporting the wind direction and their own position, and scattered reports from meteorological observers wide distances apart, making their observations at different times and with instruments of varying reliability. The most unreliable instrument, of course, was the human eye, for the weather vane, for example, does not point steadily in the direction of the wind; it fluctuates about the dominant direction and sudden gusts may throw it into an entirely different quarter momentarily. With the instruments available at that time, therefore, the exact direction of the wind, was an inference based on a more or less impressionistic judgment. With an entirely reasonable allowance for error, and for the differing times of the reports, either pattern would fit the facts. Clearly, what was needed here was a well-contrived *experimentum crucis* that could discount the possibility of human or instrumental error.

Recalling his earlier success with terrestrial magnetism, Loomis turned to the same methods—collecting all the data possible for a single storm—which he did for the storm that affected most of the United States on December 20, 1836. Loomis' observations of this storm were read before the American Philosophical Society in March, 1840, but they proved inconclusive as answers to the main question, tending to support some elements of Redfield's theory and some of Espy's. He was beset with the same difficulties that had plagued other investigators. His correspondents did not observe uniform methods of reporting—some stations, for example, did not report barometric pressure, others neglected to measure the rainfall, still others did not include the time of observation—but in spite of this he was still able to draw a number of generalizations from the data. He noted that the range of the barometer (from maximum to minimum) generally increased with the latitude, that as the barometer rose the thermometer fell, and that at the beginning of the storm in each locale there was a sudden reversal in the direction of the wind. With these generalizations, and a few other observations, he searched for a *vera causa* for each phenomenon associated with the storm.

Loomis' general procedure can be illustrated by his attempt to explain the oscillations of the barometer during the storm. The method he used was to comb the literature for all the causes that previously had been suggested for barometric oscillation. Loomis found ten of these, and he dealt with them one at a time, asking first whether or not they existed, and second, whether they were adequate to produce the recorded effect according to established laws. The oscillations had, for example, been ascribed to the destruction of large masses of air in the higher regions by electricity. Disposing of this possibility was quite easy: "The supposition," he said, "is too gratuitous to deserve serious consideration." [15] Clearly,

Loomis thought, this could not be a *vera causa* because such action was not independently known to have occurred in the upper reaches of the atmosphere.

Other possibilities deserved more serious consideration. For example, it had been suggested that the diminished pressure was the result of a whirlwind, and it had been alleged that the loss of rain could produce a fall in the barometer. In the storms Loomis was considering there had been no whirlwind, but there had of course been rain. Further calculations, however, forced Loomis to dismiss the possibility that rain was a *vera causa,* for the average amount of rain reported could have been balanced by a column of mercury about 1/15 of an inch in height—a great deal less than the range of the barometer.[16]

Another possibility was that sudden changes of temperature could have produced the effects. It was true, Loomis said, that the air over nearly the whole of the United States had become unusually heated on December 20, 1836, and its specific gravity correspondingly diminished. But although a fall of the barometer was usually accompanied by an elevation of temperature, Loomis knew that it was not always the case; in fact, he had reports which indicated that in Europe during the time of the storm a fall of the barometer had been accompanied by a fall in temperature. This indicated, "that a change in the specific gravity of the air produces only a secondary effect on the oscillations of the barometer." [17] In this case, Loomis had considered, but dismissed, the explanation now considered correct for the type of storm he was studying. His error is a particularly instructive example of nineteenth-century scientists' demand for universality. Since the relationship did not always hold, Loomis thought it could not be a *vera causa.* Not having the concept of "fronts" available to him, Loomis could only think that a storm was a storm and, consequently one contrary instance was sufficient to invalidate a hypothesis.

After thus rejecting eight possibilities on the grounds that, for one reason or another, they did not meet the test for a *vera causa,* Loomis admitted two independently acting causes which, he thought, could together explain the entire range of barometric activity. The wind blowing upward or downward near the center of the storm, the existence of which he had previously proved by similar reasoning, explained part of the movement. For the rest, he appealed to the general law of gases (relationship between pressure and volume) and tried to show from his tables that the principle was necessarily called into effect by air movement. During the storm the southeast wind which accompanied the rain moved with an accelerated velocity; thus the particles at one extremity of the current left those of the other extremity at an increased distance and diminished pressure was the result. When the storm had passed, the reverse took place and increased pressure was the result. "The causes which I have here assigned for the oscillation of the barometer," he concluded, "appear to me to be such as are known to be true, and that they are sufficient to account for the phenomena." [18]

But despite the large amount of data Loomis collected, he was still unable to deduce a certain answer to the main question.[19] He decided then that success demanded that he wait for another storm so that he could be on the scene himself to collect still more data. In February 1842 his chance came. A tornado passed over northeastern Ohio and Professor Loomis with a colleague started out for the examination of its track. The investigators were favored with a great deal of good fortune, for the tornado passed over a piece of woods near the University and hence Loomis thought the motion of the wind in passing might be inferred from the position of the fallen trees. To add to their good fortune, a second storm of great intensity was experienced in the same month, and they were able to repeat the observations.

Many others before Loomis had tried to infer the motion of the winds in a storm by observing the destruction; this was, in fact, exactly what Redfield and Espy had done in collating the reports from ships. But by this time Loomis realized that random observation of trees would give no better results than random observation of ships. If one merely plotted the angles of all the fallen trees in a devastated section of woods, he would see nothing but a mass of trees pointing in every direction whose pattern could be explained on the hypothesis of either Redfield or Espy. Instead of relying upon such random observations, he had to design an *experimentum crucis* which would once and for all decide between these two possibilities. What he had to do, Loomis decided, was to contrive some way to determine the order in which the trees in a given area had fallen. With this end in mind, he determined to take the bearing of only those trees which had fallen across each other—in such cases, the bottom one had quite indisputably fallen first and the top one last. "Having thus discovered the true *experimentum crucis* for analyzing the phenomena of hurricanes," Loomis plotted his results and found that of any given group of fallen trees, the bottom tree invariably pointed to the west and the top tree to the east. The directions of the intervening trees were intermediate between the extremes. From this generalization, the conclusion, he thought, inevitably followed: the winds of a tornado had a whirling counter-clockwise action— Redfield's theory was the correct one.[20]

Loomis published his observations of the fallen trees in the *American Journal of Science and Arts,* and began at once amassing all the facts he could find from other observers of the storm. He collected a great number of reports from the state legislatures of New York and Pennsylvania, The American Philosophical Society, the Franklin Institute, and personal correspondents all over the country. Very soon it became clear that Loomis had prematurely reached his conclusion. Even Redfield, whose theory Loomis' conclusions had supported, had to admit that the observations of trees which had fallen across each other were entirely compatible with either theory. Fresh data became, as it often does, an embarrassment as well as a blessing. Clearly what was needed was a new look at the whole question of storms.[20a] In May, 1843, Loomis sent a paper to the American Philosophical Society embodying the results of his new study. Alexander Dallas Bache read the paper at the centennial meeting the same month, and as Bache reported, it created a great sensation. It was considered then, as it is today, one of the greatest contributions made to meteorology.[21]

Elias Loomis is justly celebrated for his pioneer work in meteorology, but the irony is that he is not celebrated for his purely empirical demonstration, which is all that he thought science was about, but for the completely new method he had used. The fact is that Loomis had possessed the data to answer the questions after the 1836 storms, as had a number of other investigators, including Redfield and Espy themselves, Dove and Robert Hare; but as it turned out, having the data was not enough. It was a new theoretical conception, resulting in a new way of graphic representation, that led to the final settlement of the question. The method he had used earlier, and the method used by Espy and others, was entirely unsuited to answer any of the major questions which meteorologists were asking. Some of those questions were stated in the circulars issued by the joint committee of the American Philosophical Society and the Franklin Institute: What are the phases of the great storms of rain and snow that traverse our continent? What are their shape and size? In what direction, and with what velocity, do their centers move along the surface of the earth? Are they round, or oblong, or irregular in shape? Do they move in different directions in different seasons of the year? [22] All of these questions, as well as the difference between Redfield and Espy, appear to be of the kind that would be perfectly amenable to a purely Baconian methodology. But the resemblance to the question about magnetic declination is illusory; for one could look directly at a magnetic needle and report unequivocally what the declination was at that time and place. But he could not, in the same way, look at the path of winds in a storm. Instead, he had to survey the effects of the storm and then make a judgment about the path of the winds. As long as the decision was inferential, a number of alternative explanations were possible; thus, Loomis could not prove his point by merely surveying the damage.

The element of the argument that won Loomis his point was his graphic representation of the reports he had received. His representation on the map of the storm of 1836 had been a series of lines drawn joining the places where at a given hour the barometer was at its lowest point. That line, so far as the barometer was concerned, would mark for that hour the central line of the storm. The progress of the line from hour to hour on the map showed how the storm had traveled. This was essentially the method of representation used by Espy, and it was what made most of his observations quite worthless. There was simply no way to get answers of the kind desired from a map of that type.

But in discussing the storms of 1842, for a reason that he never explained, instead of the line of minimum depression of the barometer, Loomis drew on the map a series of lines of equal barometric pressure, or rather of equal deviations from the normal average pressure for each place. Although one can only guess, Loomis probably conceived this idea because of his earlier work on magnetic variation and dip—for this was the way these phenomena were customarily charted. At any rate, its application to meteorology was an instant success. A series of maps representing the successive intervals of twelve hours was thus constructed, upon each of which was drawn a line through all places where the barometer stood at its normal or average height. A second line was drawn through all places where the barometer stood 2/10 of an inch below the normal; and other lines at further intervals of 2/10. Thus, instead of lines of the lowest point, the deviations of the barometric pressure from the normal were made prominent and all other phenomena of the storm were regarded as related to those barometric lines. Beginning with this basis, vectors could be drawn showing the direction and intensity of the storm at any time; areas of rain or snow, of clear or overcast sky, could be represented by shadings, and the temperature could be given. In a word, with this new basis for representation—beginning with the norm instead of the lowest point—the questions could be answered.

If we are to accept his words at the close of the paper—a paper that could be called epoch-making in meteorology—Loomis had no idea that he had introduced something new into the discussion. It was, in fact, the first time the concept of deviations from a norm had been used in meteorology, and it was this concept that made the paper "epoch making"; for no significant conceptual changes in graphic-illustration have been made since his time. But in Loomis' opinion, he had merely introduced a more accurate way of representing his Baconian observations.[23]

Elias Loomis was a careful observer and a meticulous worker; his contributions to science were notable. He represents nineteenth-century Baconianism at its very best; but at the same time, in his refusal to "go beyond the facts," and in his unwillingness to even recognize that he had done more than observe and record, he also illustrates the serious weakness of that approach to science. The clear message one gets from Loomis' example is "observe, record, and one day the general laws will make themselves known to us."

According to Loomis' eulogist before the National Academy of Science, "To establish laws which had been already formulated by others, but which still needed confirmation, was to him equally important with the formulation and proof of laws entirely new." And for such a purpose, the Baconian methodology involving the *vera causa* and the *experimentum crucis, of* which Loomis was the outstanding representative, was an admirable instrument. To the question, "How does the magnetic declination vary from place to place in the United States?" one had only to collect enough determinations, place them on a chart, and the answer was there. The question, "Are the winds in a tornado circular or do they blow inward toward the center?" required a great deal more thought, but still one had only to make the right observations, arrange them correctly, and the answer appeared on the chart.

But Loomis was less fortunate in his attempted explanations. His observations could show that a proposed explanation was not the correct one: Can a whirlwind effect explain fluctuations in a barometer? There was no whirlwind in the storm of 1836 and the barometer fluctuated widely; therefore, the whirlwind is not a *vera causa*. Could rain be the cause? According to established laws, the amount of rain that fell could be balanced by 1/15 inch of mercury, the barometer fluctuated over two inches; therefore, the rain is not a *vera causa*. But when no established law fit the facts Loomis could get no help from his methodology, and the result was sometimes grotesque, as witness the appeal to Boyle's law. The same eulogist's comment on Loomis' last meteorological work could also stand as an epithet to nineteenth-century Baconianism. This work, he said, "will for a long time to come be the basis of facts by which writers in theoretical meteorology must test their formulas." [24]

As Loomis' work was for the most part squarely in keeping with the Baconian methodology, his hopes for the future of the method were entirely so:

It appears to me, that if the course of investigation adopted with respect to the two storms of February, 1842, were systematically pursued, we should soon have some settled principles in meteorology. If we could be furnished with two meteorological charts of the United States, daily, for one year, charts showing the state of the barometer, thermometer, winds, sky, &c., for every part of the country, it would settle forever the laws of storms. No false theory could stand against such an array of testimony. . . . Morever, the subject would be well nigh exhausted. But one year's observation would be needed; the storms of one year are probably but a repetition of those of the preceding.[25]

In such a context of absolute faith in the established methodology, any efforts to formulate a philosophy of science were doomed in advance either to produce "reformulations" of the philosophy of Bacon, such as that which Tyler produced with the aid of Scottish common-sense realism and Protestant dogma or to result in absurdities. If science is only observation and classification, then there is little more to be said on the subject, except by way of elaboration. The elaboration, however, as was exemplified in the case of Tyler, could often produce interesting insights into some of the other cherished beliefs of the period. Tyler and Loomis, the philosopher and the practicing scientist, represent a clear majority opinion during the early nineteenth century in America. This is not to say that they were unanimously backed, or that those who agreed with them verbally were always consistent. On the contrary, a great number of scientists were beginning to encounter the deficiencies in Baconian methodology and to try to overcome these difficulties in different ways.

❧❧❧

CHAPTER V

A Deluge of Facts

The chief source of difficulty that Baconian science faced in the early years of the nineteenth century was, paradoxically enough, spectacular success in the area of its competence, the collection of facts. Early fact gathering, being largely random and undirected, had resulted in an overwhelming mass of undigested data that ultimately became a source of embarrassment and confusion.[1] In one science after another, the older systems of data management had proved to be either unworkable or misleading, and they had not as yet been replaced with better systems or theoretical constructs. The new classification systems of the century, as observed in an earlier chapter, served the function of creating an esoteric body of knowledge and thereby providing a need for a body of experts to cultivate the knowledge. But quite apart from any professional considerations, the chaotic state of the sciences in the first half of the century demanded that some revisions be made; and the ever increasing amount of natural history knowledge that was accumulated made that necessity even more urgent. Since Baconian philosophy had been identified with classification, it was only natural that many thinkers would turn to a search for a better system of classification as the surest way out of their difficulties.

Despite its demonstrable value, there were many dangers inherent in too great an emphasis on classification and nomenclature. John D. Godman, American physiologist, naturalist, and scientific journalist, was particularly disturbed about one of the most obvious of these dangers. "Beginners of the study of natural history are generally liable to form erroneous conclusions," he cautioned, "among which none is more common and prejudicial than that of mistaking the system of classification for the subjects classed, or in other words, the arrangement of the names for the things themselves; nomenclature for natural history." [2]

Godman's recognition that there was more to science than classification—that to name a thing and place it in a structured system was not the same as to know the thing in a scientific sense—was not unique, but it did place him out of the mainstream of American scientific thought of the period. Most scientists tended to think that *when* a piece of data was finally located in its place in the truly "natural system" they would have complete scientific knowledge of that data. Bacon himself had placed great emphasis upon arranging data systematically in tables, and later Baconians were even more enthusiastic about the possibilities of such an arrangement than the master had been. Elias Loomis and other American investigators had demonstrated the impressive possibilities in systematic tabular arrangement, and their fellow scientists, sharing the optimistic outlook of the period, were far more likely to learn from their successes than from their failures. The enthusiasm was such, in fact, that one American writer was moved to attempt a natural classification of all human knowledge, including in his 581 pages, as he put it, an "exhaustive analysis" in which even the minor fragments of knowledge would find their place.[3]

Certainly Bacon had been correct in his emphasis upon the need for accurate classification, and since the end of the eighteenth century there had been a great many more reasons for systematic arrangement than there had been in Bacon's time. In his classic paper on combustion, Lavoisier had stated the need very well:

Dangerous though the spirit of systems is in physical science, it is equally to be feared lest piling up without any order too great a store of experiments may obscure instead of illuminating the science: lest one thereby make access difficult to those who present themselves at the threshold; lest, in a word, there be obtained as the reward of long and painful efforts nothing but disorder and confusion. Facts, observations, experiments, are the materials of a great edifice. But in assembling them, we must not encumber our science. We must, on the contrary, devote ourselves to classifying them, to distinguishing which belong to each other, to each part of the whole to which they pertain.

The French chemist was aware of the dangers inherent in unsystematic collection of data. His new chemical nomenclature was designed to provide a ready way to classify all the materials of chemistry and assign them a place in the system. Furthermore, it contained a built-in device for determining what to do with new substances as soon as they were discovered; for without this feature he realized that each new fact, rather than contributing to the progress of the science, would actually be an encumbrance.

The system of Lavoisier, for a while at least, had cleared up the chaotic state of chemistry and reduced it to order. Prior to the work of the French reformers there had been no systematic way of naming substances, and many had retained names given them in classical antiquity, or by medieval alchemists, or simply by common usage. Under such conditions it was difficult to master the names of the substances, let alone understand their behavior. Substances had been named from their place of discovery or occurrence (Epsom salt), or after their discoverer (fuming liquor of Libavius), or from analogy to a familiar substance (butter of arsenic, butter of antimony, flowers of zinc). Such names as these, Lavoisier held, were not only ridiculous but inimical to the progress of the science, for they often gave rise to false ideas. Most of the "butters," "flowers," and other innocent-sounding substances, Lavoisier pointed out, were in reality deadly poisons, bearing no chemical relationship to the substances from which they derived their names. In place of this, Lavoisier and his associates proposed that all substances be assigned a chemical name which would indicate the elements they were composed of and the manner in which they were combined. In short, a scientific name should be functional; it should describe the chemical composition as precisely as possible. In order to achieve this object, Lavoisier had constructed tables, showing in the first column the simple substances, divided into five classes; then in another column compound substances were listed, with the names indicating the simple substances of which they were composed. The acids, for example, formed one class of compounds. Oxygen, the "acidifying principle," was thought to be common to them all, and the other element varied from acid to acid. They were therefore placed in the column next to oxygen and the individual names were formed from the varying substances. The acid produced by the combination of sulphur with oxygen was named sulphuric acid, and a similar acid containing less oxygen was named sulphurous acid. Their salts became, respectively, sulphates and sulphites. The same principle was followed when the bodies composed of three simple substances were named. "We have to consider," he said, in the bodies that form this class, such as, for example, the neutral salts, firstly, the acidifying principle that is common to all, secondly, the acidifiable principle that constitutes their particular acid and, thirdly, the saline, earthy, or metallic, base that determines the particular kind of salt. We have taken the name of each class of salt from that of the acidifiable principle common to all the members of that class and we have then distinguished each kind by the name of its particular saline, earthy, or metallic base.[4]

But with the breakdown of Lavoisier's neat arrangement, caused in part by the discovery of new elements that stubbornly refused to fit into his tables, and in part by the discovery that the operations of nature were not so simple as the great French chemist had thought, there was nothing immediately available to take its place. Briefly, Lavoisier's system had been founded on the relation that oxygen was supposed to bear to the other elementary bodies; it was believed to be the only supporter of combustion and the principle upon which the acids depended for their peculiar properties. Shortly after Lavoisier's death, however, a number of acids were found which contained no oxygen; and by the second decade of the nineteenth century at least three other substances were generally classed with oxygen as supportrs of combustion (chlorine, iodine, fluorine). In fact, as one American investigator observed with some perplexity, the distinction between combustibles and supporters of combustion did not even seem to be a tenable one.[5]

Although American commentators were certainly exaggerating when they maintained that Lavoisier's system had been "overthrown" by modern discoveries—for his principles are those upon which modern chemistry is founded—yet, it had lost a great deal of its appeal as well as its utility when oxygen had to be removed from its dominant place. Since it was the assumed necessary presence of oxygen in those two chemical operations that made the system so comprehensive and so psychologically satisfying, American chemists felt that in removing this element from its position of dominance all the certainty and all the security had been abandoned.

It is against this background of the recent loss of what was commonly thought to have been certainty—the loss of the idea that each subtance must have its own unique place in the system, being responsible for an operation that no other substance could perform—that American reactions to the accumulation of chemical knowledge must be understood. Nature had seemed so much more regular and so much more orderly—more worthy of the Creator responsible for it—when things had been simpler. When John Ware, Boston physician and naturalist, observed that he was "astonished at the accumulation of facts which has been made within the present century," and that the "part of the volume before us which relates to the details of the science, seems almost wholly occupied in the narration of new discoveries," he was not unequivocally praising chemistry. He was , in fact, registering a definite complaint, for "the science consists now in an immense number of facts without any regular and consistent connexion." [6]

Benjamin Silliman, reviewing Thomson's work on organic chemistry, made much the same complaint, remarking that the time had already arrived when all the facts of chemistry could not be taught within the limits of any college course or sequence of courses. And he struck a sobering note, which was at the same time quite prophetic of the future, when he observed that "if the full detail of facts is still to be given, the science must be divided under different professors and comprised in several courses constituting a system." Silliman hoped for the day when the science could be taught by means of a few general principles, but at the time he wrote he had none to offer.[7]

Scientists had, at some point in the opening decades of the nineteenth century, reached the limits of their Baconian method. And with the observed inadequacy of Lavoisier's heroic effort it became painfully evident that this was so. More than any other factor, it was this simple accretion of knowledge, the heaping up of fact upon unrelated fact, that was responsible for the exhaustion of the method's potentialities. Scientists had collected so much data that it had become unmanageable, and the failure of the data to arrange itself automatically was becoming more and more obvious with each new discovery. Science was supposed to begin with natural history and move gradually, with little human assistance and no imagination, into ever higher levels of generalization, but natural history was showing no indication of ever leading beyond itself. The *Philosophia Prima* seemed even further away to the scientists of the early nineteenth century than it had to those of Bacon's own time, who had been confidently informed that a single generation would suffice to give man a complete knowledge of nature.

As Denison Olmsted observed in a review of "the present state of chemical science" (1826–1827), the difficulty Bacon had failed to anticipate was that the logic of science might not keep up with the knowledge of it. During the past fifteen years the progress of the science of chemistry had been so rapid that, according to Olmsted's report, elder scholars frequently complained that it had passed beyond their capacity. But the vast accumulation of knowledge was almost exclusively empirical; when chemists tried to account for any of the great number of facts they knew they found that they could assign no adequate reason for any chemical action. All they could do in this situation was "arrange such phenomena of this kind as are analogous into separate classes." To explain chemical actions by reference to general laws seemed quite unsatisfactory to Olmsted, for this was only to make a tautology.[8]

An anonymous Boston physician spoke in much the same manner about his own field of science. "Whoever will look over the medical journals of the last fifty years," he complained,

will find them filled with facts at variance with each other and flatly contradicting every general law laid down in the materia medica; and he who should attempt to draw out from the mass a limited number of primary effects of the various articles to which all the rest might be referred, would soon be convinced that such a task is wholly impracticable.

In comparing the state of medical knowledge in his own day with that of the ancients, the disillusioned physician observed that any superiority the moderns might have was not due to discoveries of new general principles. It was, he concluded, that "the discovery of new facts, and the application of a more rigorous plan of examination, had enabled us to reduce the number of those once supposed to be established." Instead of understanding the causes of disease, physicians had spent most of their time mastering the names of an unbelievable number of presumed specific disease entities. With a true Baconian faith that understanding would be the automatic outcome of precise distinctions, medical investigators had discovered upward of two thousand "species" of disease, and on the basis of such entities had erected an imposing superstructure consisting of orders, classes, and the like. But their expectations had been disappointed. In the second quarter of the nineteenth century, physicians were still treating symptoms, not diseases. Miraculously, some patients continued to recover, but physicians seemed no closer to understanding than they had ever been.[9]

Since it was widely understood that classification was a necessary step in making information manageable—even by that minority which did not regard it as the whole of science—many scientists during the early nineteenth century were concerned in a major way with arranging the data of their respective sciences. As William Whewell pointed out, classification must be assumed before one could even begin his researches.[10] The only difficulty was, as Lavoisier's failure had demonstrated, that of discovering a system that could provide the unambiguous determinations required, and which could maintain sufficient "openness" to accommodate new discoveries.

James Dwight Dana, with what was to prove an unwarranted optimism, thought that for mineralogy such a system had been found and all the problems of nomenclature solved by 1836. In a paper read before the Lyceum of Natural History of New York, he advanced the opinion that the chief obstacles to the introduction of a systematic method of naming the mineral species had gradually been removed during the rapid progress which had occurred in the past few years. Specifically, Dana endorsed the system of the German mineralogist Mohs, and although he felt it necessary to make a bow to chemistry by observing that Mohs had been guided to a certain extent by "that important source of physical characters, viz. chemical composition," he was pleased to announce that Mohs had not wholly relied upon chemical characters; "the exclusive adoption of which would have degraded mineralogy from the rank of an independent science, and merged it in that of chemistry." [11]

Although Dana made the required bow to chemistry, Mohs' nomenclature was really of the natural-history type, based upon the same external characteristics that had been in vogue since the time of Werner—lustre, hardness, specific gravity, color, and others; as witness Dana's description of a mineral cabinet arranged according to the system of Mohs:

The Gases and Liquids, with which the arrangement commences, are followed by the Salts, so disposed as to present an increase in stability, hardness, and lustre, as the eye proceeds onward. Among the Gems, we arrive at the diamond, in which these characters reach their climax.

Thence descending in the series, we gradually pass through the Metallic oxyds to the Native metals. In these, the light-coloured species are followed by the Sulphurets and Arsenids of similar color and lustre, which are succeeded by the dark-coloured metallic sulphurets; and these pass insensibly to the sulphurets without a true metallic lustre. From the latter there is a natural transition to Sulphur, and its close allies the Resins and Coals, with which the series terminates. By this association of species agreeing in external characters, the attention of the student is naturally led from the observation of their many resemblances, to a particular consideration of their several peculiarities.

Dana was sure that this arrangement displayed "a chain of affinities running through the whole, and connecting all the several parts." [12] Although he was to change his mind less than ten years later and become one of the leading champions of chemical nomenclature, his early objection to basing mineral classification upon chemical characteristics was a typical response to the encroachment of chemistry upon natural history that was taking place at the time.

The encroachment had begun quite early in mineralogy, and as usually happens, the early supporters of chemistry had been dogmatic in their rejection of the older method. As early as 1817 the argument was uncompromisingly stated by a writer in the *North American Review*. The great difficulty in the past, he said, had been the absence of a "fixed and invariable point," from which to begin in the determination of mineral species. As in the organic world, species were determined by such fixed characters, transmitted from one individual to another, so it must be in the mineral kingdom. What, then, constitutes this fixed and invariable character? he asked:

Colour does not; nor form, hardness, transparency, or any other external mark. These are all variable, and nothing in minerals is determinate but their chemical composition. All, which possess the same composition, belong to the same species; and it is a knowledge of this only, which can lead to proper specifick distinctions. [13]

Knowing nothing of such complicating factors as isomerism and isomorphism, the writer was confident that chemistry had, through advances in electro-chemical analysis and through the theory of definite proportions, progressed to the state where it could yield exact determinations. John Ware, writing in the same journal two years later, was not so confident. "The forms of matter are always the same," Ware argued, "although our views of their constitution may be constantly changing." The relative stability of forms meant to Ware that a system erected on natural-history principles would be immune to all the "confusion and disorder" that the current changes in chemical theory were producing.[14]

In 1837 James D. Dana had published the first edition of his *System of Mineralogy,* using the natural history method of classification, complete with Latin nomenclature, classes, orders, and genera. In the 1844 edition he preserved the same arrangement, but appended a supplement in which a classification based upon chemical principles was proposed, and which he subsequently adopted in the edition of 1850. After Dana's conversion to chemical nomenclature in 1844, he joined Benjamin Silliman Jr. in his effort to purge the profession of the natural-history method by using the influence of the *American Journal of Science and Arts.* The younger Silliman began his efforts in 1844, the year of Dana's conversion, by attacking a work of Charles U. Shepard, former student of his father and America's leading natural history oriented mineralogist. "Gravitation is not more essentially connected with astronomy, than chemistry with mineralogy," Silliman announced, "and the student who is ignorant of the composition of his minerals, might with equal profit possess a cabinet of glass of divers hues." [15] Shepard, on the other hand, clung to his belief that mineralogy was a branch of natural history, the sole function of which was to illustrate the physical properties of minerals, and "thereupon to name, classify, distinguish and describe the species." He did, Shepard granted, sometimes make chemical analyses, "but never with the idea that I am mineralogically employed." [16]

Although Silliman severely censured Shepard for not admitting chemical evidence in his classification, he was, nevertheless, openminded enough to grant that there were certain difficulties even if chemistry were used as a basis for classification. For example, as far as the chemical analyst could tell, rectile and anatase, Sillimanite and Ryanite, white iron and iron pyrites, carbonate of lime and arragonite, garnet and idocrase, were identical. Yet, in their appearances they were evidently distinct. And Silliman understood as well as Whewell did that no system of classification could be considered adequate if it did not account for physical differences. This lack of a necessary correspondence between chemical and physical characteristics, at a time before dimorphism and isomorphism were well understood, posed a grave problem for those who wanted to argue for classification by either method. Silliman, after making the admission that chemistry alone was not adequate, closed by looking forward to a new method for mineralogy "which should combine all that was essential and truly important, both in chemistry and natural history, without a blind adherence to either." [17]

In addition to the fear for the position of mineralogy as an independent science, and the general recognition of chemistry's equivocal success, some thinkers also felt that making mineralogy an appendage to chemistry might lead to unfavorable results for the latter science. There was no doubt, as William Whewell said, that the elementary composition of bodies must determine their properties. This meant that all mineralogical arrangements, whether they professed to be or not, must be, in effect, chemical. The object of any arrangement must be to display chemical relations, but at the same time, he thought that the classification must be made on some other basis if it were to be of any value to chemistry. "If chemistry be called upon to supply the *definitions* as well as the *doctrines* of mineralogy," he said, "the science can only consist of identical propositions.[18]

Whewell, viewing the problem from the position of the general philosophy of science, was afraid that chemistry might become no more than a system of tautologies if there were not some independent basis for naming the things being operated upon. There were those who believed that this unfortunate state of affairs had already, in part at least, arrived. Robert Hare saw this tendency in the chemical nomenclature of Berzelius:

While an acid is defined to be a compound capable of forming a salt with a base, a base is defined to be compound, that will form a salt with an acid. Yet a salt is to be recognized as such, by being a compound of the acid and the base, of which, as I have stated, it is made an essential means of recognition.[19]

Besides the tautologous nature of the system, Hare objected to Berzelius' nomenclature because it would exclude from the class of salts substances which the mass of mankind would still consider as belonging to it; for example, sodium chloride, or common table salt; and it would assemble under one name, combinations opposite in their properties and having none of the qualities usually considered indispensable to that class. If every compound of an acid and a base were to be considered a salt, he pointed out, such substances as marble, glass, and porcelain would have to be on the list. Hare's solution would combine an electrical decomposition test with a chemical test and a natural-history test:

> When two compounds capable of combining with each other to form a tertium quid, have an ingredient common to both, and one of the compounds prefers the positive, the other the negative pole of the voltaic series, we must deem the former an acid, and the latter a base. And again, all compounds having a sour taste, or which redden litmus, should be deemed acids in obedience to usage.[20]

Hare's general views on the nature of matter—that it was a theoretic inference from our perception of its properties—gave him no satisfactory way to distinguish between "properties" and "qualities"; he perhaps lacked the psychological sophistication to make the distinction on other grounds. At any rate, his admission of "sourness" independently of any chemical test was a survival from the older view of "principles," and it made his otherwise rigorous chemical formulation essentially a natural-history one.

According to Whewell, efforts at scientific classification had been a great deal less successful in the sciences dealing with inorganic substances than in those dealing with organized beings. This was so primarily, he thought, because the necessity for such a systematic effort had only recently come to be sufficiently recognized by chemists.[21]

Most American scientists, however, were not so certain as Whewell that entire success had marked the efforts of natural historians. Asa Gray, reviewing John Lindley's *Natural System of Botany*, quoted with agreement Lindley's statement of his fears for the future of systematic botany. The advancing knowledge of botany and the discovery of new structures, he said, made him fearful that "another chaos" would be brought on by the masses of imperfectly grouped species with which the science would soon abound. Gray noted that Lindley listed 7,840 genera in his enumeration, published in 1837. Ten years earlier, Sprengel had been able to list only 3,593 genera, or not quite half the number known to Lindley. The twelfth edition of *Systema Natura* (the last of Linnaeus himself) contained only 1,228 genera.[22] Linnaeus' admittedly artificial system had brought a measure of order out of the earlier chaos in botany and zoology, but he had himself pointed to the inadequacy of his own system arising out of its lack of a truly natural basis. The "Prince of Botanists" had not only enjoined his students to dedicate themselves to a search for the truly natural system, but had begun the work himself, publishing sixty-seven orders which he thought were "natural."

During the first half-century after Linnaeus' death, however, botanists (with the exception of a few in France) had remained so enamored of the artificial sexual system that they had generally forgotten the master's injunction. In Gray's day the inadequacy of artificial systems was felt a great deal more severely than in Linnaeus' time. By the beginning of the nineteenth century, as thousands of new plant forms were being brought in from Africa, North America, the West Indies, and other areas new to European science, it was becoming increasingly evident that the Linnaean catalogue was of quite limited utility. But with all the progress in the collection of data, no acceptable systematization on a natural basis had yet been developed—although a great many rival systems had been proposed. In the single twenty-year period from 1825 to 1845 at least twenty different "natural systems," not including the great number inspired by *Naturphilosophie,* had been published.[23] Yet, to Gray, writing in the middle of this great era of botanical system building, it appeared that the chaos had only become worse. All the proposed classifications began with an a priori judgment (although rarely recognized as such) of which organ system was "most important." And in every system some group or other formed in accordance with it was demonstrably unnatural.

In part, the proliferation of species was due to new discoveries, many of which were made on the exploring trips into the interior of the American continent. Some increase in the number of known species could therefore be expected as a consequence of the normal growth of natural-history knowledge. But an even more important —and much less desirable—source of new species was the growing tendency to further subdivide already determined species or else to place them in new genera or otherwise regroup them, often for no apparent reason. The widespread acceptance of the Linnaean system, which directed the student single-mindedly toward the mechanics of finding the name of an object, had predisposed naturalists to look more and more at their material from a single point of view—to direct all of their attention toward the single goal of finding the name of a thing. And in their anxiety to facilitate the finding of the proper name, both botanists and zoologists seemed to have developed a positive mania for "species-splitting," as the practice was called. The attitude even carried over into the era of natural systems. As one American naturalist expressed it:

Habits, uses and stations of insects in the system of nature have been neglected for the acquisition and description of species, the indication of new genera, and the coining of an immense number of new and pedantic terms, in a science whose nomenclature is already overburdened.[24]

Protests like that of the American naturalist came from all sides, but given the state of natural-history knowledge at the time, and the lack of any generally agreed-upon standards, there was little that could be done about it.

After John Torrey's introduction of the natural system to America in 1826, followed by L. C. Beck's and Asa Gray's early advocacy, American botanists for the most part joined in the search for the truly "natural" system which would be an authentic representation of nature's family relationships.[25] Zoologists, too, joined in the search in such numbers that by 1834 the Reverend Leonard Jenyns, speaking before the British Association for the Advancement of Science, was able to observe that "it is now generally acknowledged that the true and legitimate object of zoology is the attainment of the natural system." [26]

Despite the in-principle agreement, a major difficulty of the systematizers was that it was never made entirely clear what was meant by a "natural system," except that it was founded on "true affinities" and "indicated by nature itself." In the absence of any other generally agreed-upon principles, scientists could do no more than caution each other that in arriving at their classification systems they should be careful to "copy that arrangement . . . which nature itself has made." [27] When the natural system was attained, however, it was generally assumed that it would be based upon over-all structural and functional resemblance. The problem of species, to put it very briefly, was always stated in terms of the "characters in common" criterion, for it seemed axiomatic that those organisms were most closely related which shared the most characteristics, and, conversely, those were most distantly related which shared the fewest characteristics. To those schooled in the additive nature of scientific investigation, it seemed entirely reasonable to assume that

if one only observed, recorded, and compared as many characteristics as possible, nature's own arrangement would very soon
emerge. Asa Gray, one of the leading proponents of the natural
system in America, stated its assumptions as follows:

That the affinities of plants may be determined by a consideration of all
the points of resemblance between their various parts, properties, and
qualities; that thence an arrangement may be deduced in which those
species will be placed next each other which have the greatest degree of
relationship; and that consequently the quality or structure of an imperfectly known plant may be determined by those of another which is well
known. Hence arises its superiority over arbitrary or artificial systems,
such as that of Linnaeus, in which there is no combination of ideas, but
which are mere collections of isolated facts, not having any distinct
relation to each other.[28]

Despite Gray's deprecation of "mere collections of isolated facts,"
the Baconianism implicit in his own view of classification at that
time is inescapable. One had merely to consider and compare
enough facts and the proper "natural" arrangement could then be
"deduced" from this knowledge—in exactly the same manner that
theories were to be "deduced" from facts in the physical sciences. A
belief like this, at times, resulted in efforts that a later generation
judged to be grotesque. The extreme example is probably Michel
Adanson, who framed sixty-five different systems of botanical
classification, each based upon a single mark. His supposition was
that natural affinities would emerge as the ultimate product.[29]

Despite the comprehensiveness of the ideal, actual efforts at
natural classification of plants and animals were invariably based
upon something less than a comparison of all the parts of an
organism. With the vast number of possibilities each organism
presented, some compromise was necessary. As usually happens in
such an attempt, the seemingly reasonable effort to classify things in
terms of their real natures opened up the gateway for a whole
philosophy of nature to enter a classification system. Efforts were
based upon a comparison of what the researcher in question
thought to be the most important organs. Although the judgment of
what was "most important" was bound to vary from system to
system, in accordance with the pervasive teleological assumptions of
that period, it was generally agreed that comparison either with
reference to those functions "destined to the preservation of the
species," or those "destined to the preservation of the individual"

would yield a natural classification. In practice, this meant that prominent or striking features, and those that were "connected with the general economy" of the organism, were assumed to be the most important for purposes of classification, and other characters were considered subordinate to these. The taxonomy of the time, in fact, directed that one not be concerned about differences of structure which were not accompanied by corresponding differences in the rest of the organization.[30] It also meant, for most investigators, that only mature forms would be compared with each other, for only after maturity would the organs be fully adapted to the destined conditions of existence. Thus the presuppositions of the naturalists dictated that their work be done under the worst possible conditions for the attainment of their objective. The effort to classify in terms of "real natures" directed attention to the least stable of all characteristics—namely, those most connected with the style of life of the organism. And it meant that the vestigial, the abortive, and the useless—precisely those that later naturalists found of greatest significance—would have no place in determining relationships. The early proponents of the "natural system" thus fulfilled the dual function of confirming the teleological belief in "wise contrivances" and of militating against any evolutionary conclusions.

That the desired arrangement could not be mechanically deduced from a mere consideration of selected facts, or even of all the facts, was already occurring to many scientists of the period. John D. Godman, although quite unique in going to such an extreme, held that all classifications were, in the very nature of things, arbitrary. The best possible arrangement, he held, "is but a summary of distinctive epithets and characters to aid in the arrangement of knowledge." [31] A more typical attitude, showing the beginnings of disillusionment much more clearly than the nominalism of Godman, was that of Gray himself. In the same article in which he approved the principles of natural classification, Gray announced some very significant reservations as to the possibility of applying them:

This might easily be accomplished, if the idea once so strongly insisted upon by poets and metaphysicians, of a chain of beings, a regular gradation, by a single series, from the most perfect and complicated to the most simple forms of existence, had any foundation in truth. On the contrary, nothing is more evident, than that almost every order, or another group, is allied not merely to one or two, but often to several others, which are sometimes widely separate from each other; and, indeed, these several points of resemblance or affinity, are occasionally of about equal importance. A truly natural lineal arrangement is therefore impracticable, since by it only one or two out of several points of agreement can be indicated.[32]

Gray, with these words, explicitly rejected the possibility of achieving the objectives of the systematists by using their current assumptions. But he had not yet—although he did later—come to reject their assumptions. That a classification, however imperfect, might be *deduced* from a contemplation of the parts was a belief that scientists of the period were driven to because of their inability, or unwillingness, to inquire behind the affinity for the cause of the affinity. By "affinity" they merely meant "structural resemblance," and without a theory to explain the cause of the structural resemblance—or, indeed, to justify ignoring it in some cases—the problem of botanical, or zoological, taxonomy, was insoluble. The very idea of "natural" relationship itself, upon which the natural system depended, necessarily remained a great mystery to all who believed in the constancy of species. Since no scientific meaning could be attached to the concept, the only recourse was to indefinite expressions with figurative meanings. Expressions like "plan of symmetry" and "type" (in the sense of an ideal original form) drove systematists to the conclusion that in their work they were expressing the original plan of creation—representing the thoughts of the Creator Himself. Operating in such a framework, Gray's disenchanted view of the future seemed justified. All that he could see for the future of systematics was piecemeal improvement; the Baconian method had revealed such an intricate web of nature that the method seemed in danger of being strangled by it.

❧❧❧

CHAPTER VI

The Limits of Baconianism:

History and the Imponderables

In 1848 James D. Whelpley came to a conclusion that must have shocked many of his contemporaries. He noted that there was a defect in the commonly held notion that no theory could be admitted as true until it had been tested by experience. Despite the fact that Baconian philosophy was very clear about this matter, the criterion simply could not be met in some very important cases, Whelpley argued, for it was manifestly not true that any amount of experience would ever make us acquainted with the "arcana of nature." It was, to be precise, quite often the most scientifically interesting portions of that arcana that were concealed from the senses. We could never, for example, feel a particle of matter or see a pulse of light; and yet, these phenomena were of transcendent importance in virtually every scientific investigation.[1]

With these words Whelpley had clearly stated the nature of the dilemma that scientists had been facing more and more since the end of the previous century. The crude empiricism which was popularly regarded as the essence of Baconianism demanded that one confine his researches to that which was physically accessible. The grave limitations of Baconian empiricism did not become evident so long as the major scientific need was the collection and description of data. The method had, in fact, produced truly impressive results in those areas. But by the beginning of the nineteenth century, as the early data-collecting stage grew to a close in a number of sciences, the Baconian method had begun to show signs of having exhausted its potentialities. Even in the primarily classificatory sciences, as the previous chapter indicated, Baconianism had come to be an equivocal blessing. And in the other sciences it seemed even less of a blessing. By the beginning of the century scientific attention had become focused upon things which were not immediately evident; upon things that simply could not be investigated by direct observation in an unsophisticated sense of the term.[2] It was clear to many, however, that further progress depended upon finding a way to deal with them. Vague murmurings of discontent began to be heard from men who were colliding against the hard walls of their methodology, although the depth of their commitment was such that few could even understand the source of their perplexities as clearly as Whelpley. Something had to give way in such an unsatisfactory situation. And very often it was the philosophy that was stretched beyond recognition or quietly ignored, but seldom explicitly abandoned.

It was previously noted that a man like John Esten Cooke could open his book with the conventional declaration of his allegiance to Newton and Bacon and then proceed forthwith to spin a single-entity theory of disease entirely out of his own mind. In this respect Cooke was no isolated case, for despite the example of men like Elias Loomis, the practice of going significantly far "beyond the facts" was becoming more and more common during the first decades of the century. If one were to judge from the behavior of a great many scientists, he would be tempted to describe the attachment to pure Baconianism as largely verbal. When scientists at the frontiers of knowledge were not speaking directly of their accepted theory they seemed to forget that they were committed Baconians. Besides being in itself a most effective argument against the eternal validity of their method, this ambiguous situation meant that a wide gulf had already developed between the scientists' acknowledged guiding theory and the theory actually underlying their practice, or even their explanations of nature. Such ambiguity provides an added reason why efforts to formulate a philosophy of science so often ended in patent absurdities.

The closer one came to the fundamental building blocks of the universe, the greater loomed the difficulties posed by Baconianism. One could certainly not see two atoms combining with each other to form an atom [3] of a third substance; yet by assuming that this was the way chemical combinations occurred, scientists could satisfactorily "explain" the Baconian fact that elements always combined in definite proportions. And, accepting this theory of atomic combination, a really bold speculator could be led on to yet higher levels of abstraction in explaining other phenomena. However convenient this might be, the acceptance of a theory because it explains some otherwise inexplicable phenomena was wholly unlike the Baconian grounds for admitting theory. A theory, as Silliman observed, should "flow naturally as an induction" from the facts, and go no further than the facts themselves warranted.[4] That is, it was thought to be a simple summary of the type Elias Loomis had made from his meteorological tables.

Accepting a theory on the grounds of its explanatory power, in addition to being difficult to reconcile with the simple building process upon which science was thought to proceed, was also opposed to contemporary interpretations of the Newtonian *vera causa*. The meaning usually given to the *vera causa* was that discussed in connection with Loomis' work: "The Newtonian rules of philosophizing require that nothing should be assumed as the cause of any phenomena, unless it be proved adequate to their production, and also known to exist."[5] This was the rule that Loomis had applied so rigorously in his meteorological investigations where it had, in general, worked quite well. But the attempt to apply the rule with the same rigor in more abstract investigations led to grave difficulties, if not outright prohibitions. It was the second requirement that the higher levels of theory completely failed to meet, for those who appealed to this rule meant that the alleged cause must be known to exist *independently of theoretical calculations in the place and in the* quantity required *by the hypotheses.* In the view of the more demanding philosophers of the period, even evidence admitted under this interpretation of the Newtonian rule would have to be termed "merely circumstantial." Even under the most favorable circumstances, the most that one could conceivably say in favor of the atomic theory was that there was a "presumption" that it might be true; this much could be said if, and only if, the atoms were known to exist independently of any of the facts they were to be called upon to explain, "for a presumption cannot be raised on a mere conjecture."[6]

This interpretation of *vera causa,* which does great violence to Newton's meaning, was one of the ways that Newton and Bacon were brought together to form the Baconianism of the nineteenth century. A theory was thought to be the simple generalization of a fact; that is, a general statement which included a great number of facts of the same nature. In modern terminology we would say that a theory was then considered no more than a synthetic all-statement—which would now be termed a "law." [7] One investigated "enough" swans, to borrow a common example from the logicians, and then generalized that "all swans are white." The allegation that all swans were white was then thought to be a legitimate theory.[8] Or, to make the problem a trifle more difficult, but still essentially the same—so it was assumed—one observed the motion of apples and planets and then generalized that "all bodies obey the law of gravitation." The problems involved in Newton's "discovery" of the theory were thought to be simply those of discrimination and classification. The general law derived, in both cases, from a careful adding up of particulars.[9] Anything that could not be placed in the framework of induction by simple enumeration was called "hypothesis," a popular word of disapprobation at the time. Theory was thought to differ from hypothesis, as the Scottish philosopher Thomas Brown stated the distinction, only in that it did not suppose the existence of any substance that had not been actually observed, "but merely the continuance in certain new circumstances of tendencies observed in other circumstances." [10] The distinction was stated somewhat less accurately, but much more colorfully, by a reviewer in the *American Monthly Magazine and Critical Review:*

Theory may be considered the inductive process of the understanding; hypothesis the syllogistic method. Theory travels in the humble *posteriori* road from particulars to a general conclusion. Hypothesis drives along the high *priori* road, applying assumed rules to particular cases. Theory may be termed the logic of nature, hypothesis that of art.[11]

Significantly, the reviewer did admit, albeit reluctantly, in the next sentence that both theory and hypothesis were necessary.

Most American scientists were not inclined to allow hypothesis even the limited role suggested by the anonymous reviewer. The reaction of Dr. John Ware to one important effort to use "hypothesis" is indicative of the orthodox attitude. Arguing against both the material theory of heat and Davy's kinetic theory, the Boston physician dismissed both with the comment that he was "not obliged" to show how things could be managed without the existence of caloric. The burden of proof lay on the other side, and

Davy's attempt to account for the phenomena was "purely hypothetical" and besides that "somewhat obscure." As a way of dismissing the subject, Dr. Ware concluded that "We are not called upon to enter so deeply into the hidden processes of nature." [12]

In terms of these distinctions among "law," "theory," and "hypothesis," the atomic theory of chemistry was no more than an "hypothesis," for it had been derived from "applying assumed rules" to a great number of particular cases. Despite the protests of people like Ware, it was reluctantly admitted into chemistry because "if there was not a law of nature to determine and preserve these fixed proportions, there could be no uniformity in compound bodies." [13] Because of its obvious utility, the atomic theory was generally accepted, at least in its fundamentals, in the early years of the century; but it always remained suspect because of its "hypothetical" nature, and by the beginning of the 1830s had fallen into disfavor, not to gain general acceptance again until late in the century after acceptance of Avogadro's law made it possible to solve some of the difficulties of the theory.

Keeping in mind the extremely limited definition of a theory at that time—that it could be nothing but a direct result of induction by simple enumeration—one can understand this otherwise inexplicable comment by a respectable scientist on the relation between theory and experiment:

I must first premise that the present state of knowledge on the subject does not enable us to take into consideration all the circumstances which combine to constitute the resistance encountered by a body in moving through a fluid; and therefore the conclusions of theory should not be expected to coincide with the results of experiment, but should be regarded as showing simply what would be the results of experiment, if the circumstances taken into consideration were the only circumstances tending to modify the result. It is not safe therefore to advocate a theory, as one of your correspondents [L. R Gibbs.] does (Vol. 28, page 137) on the ground that its conclusions coincide more nearly with the results of experiments, than do the conclusions of other theories; for it is evident from the consideration just stated, that the theory whose results differ most widely from the results of experiment, may nevertheless be the true theory.[14]

The line of argument used by Eli Blake clearly assumes that theories are merely additive in nature. Given this definition, the argument then becomes comprehensible; we do not know all the components in the resistance of fluids; experiment involves all the components; therefore, the theory whose results differ most widely from the results of experiment may be the true one.[15]

Such a simple arithmetical idea of a proper theory was one of the first limitations that those who wanted to inquire very deeply into physical chemistry had to overcome; and as with physical chemistry, the same general comment could be made about electricity, a new science which began development about the end of the eighteenth century. No one could observe an electrical current in action—in fact, one could not by any of the ordinary rules of philosophizing show that such a thing existed. The scientist in this case was limited to observing certain effects of this hypothetically inferred entity. It was the study of entities like electricity, the so-called imponderables,[16] that proved the severest test to Baconianism during the early part of the century. To admit that there might be a matter which did not have weight—the electric fluid, caloric, the ether, light, and magnetism—was an affront to the whole basis of Baconian philosophy. As Whewell, certainly no doctrinaire Baconian in most matters, said: "if there be any kind of matter which is not heavy, the weight can no longer avail us, in any case to any extent as the measure of the quantity of matter." [17] It seemed to Whewell, as to a number of other scientists, that the admission that there might be a weightless matter would require such a drastic revision of methodological procedures that none of the generally accepted categories would thereafter be useful. He could not even imagine a quantitative measure that could replace the current weight concept —and by most scientists of the time the inability to conceive of a phenomenon was considered even more damning than the inability to see it. For Whewell, giving up the idea that all matter had weight was equivalent to giving up science. Nor did Whewell like the modification of the Newtonian requirement of *vera causa* that the study of the imponderables had necessitated. According to the revised viewpoint, that it required causes which were known to exist by their mechanical effects, the electric fluid seemed to be such a *vera causa*. An electrical current could be made to produce sparks, shocks, or explosions; and these effects were cited by defenders of the theory as proof of the reality of a material electric fluid. Whewell argued that this showed how delusive the Newtonian rule, so applied, could be; it "consists in elevating our first rude and

unscientific impressions into a supremacy over the results of calculation, generalization and systematic induction." [18]

The issue of the weightlessness of the electric fluid, although not
its reality, could be dodged, and many scientists found a convenient
way to do this by appealing to limitations in current instruments—
an appeal that has historically been among the most potent of the
conservative forces at the disposal of a scientific community. Benjamin Silliman dismissed the matter by saying: "It cannot be weighed,
because our balances and organs of sense are not sufficiently delicate." [19] Presumably, a better instrument might someday be devised, and it was wise to defer judgment until such time. There
was no a priori reason, as Denison Olmsted argued, for supposing
that nature stops in her processes of attenuation precisely at the
point where, for want of more delicate instruments or more powerful organs of sensation, our methods of investigation came to their
limit.[20] For all practical purposes this was as good a solution as any,
and the appeal to limitations of instruments had certainly been
justified by future developments many times before.

But such an appeal did not satisfy the desire displayed by people
like Robert Hare to know what the world was made of; for as long
as this appeal was automatically made in the face of every difficulty
there was no conceivable way to advance beyond the current
concepts. Hare thought that there was already more than enough
evidence to indicate the existence of imponderable substances, and
he did not shrink from the inevitable conclusion that their existence
required a radical redefinition of the concept of matter. There were
known properties in the world that could not then be ascribed to a
matter having weight and, consequently, obeying the Newtonian
laws. But according to Hare there could be no property without
matter in which it was inherent. "Nothing can have no property," he
bluntly declared.[21] A few years later, in his textbook prepared for
students at the University of Pennsylvania, he defined matter simply
as "that which has properties,"—with no precise properties being
specified as essential.[22]

A man who spoke in this manner today would be called a "Materialist," but in discussing the early nineteenth century, the word has
no meaning unless it is further defined. If there can be a matter
endowed with none of the properties ordinarily associated with
matter, the word itself ceases to convey any information. As Hare

said, the only things about which one could be certain were properties, and the possessors of the properties were inferred entities. In
one of his many arguments with Michael Faraday, Hare made it
clear in what sense he was a "materialist."

I agree that "a" representing a particle of matter and "m" representing its
properties, it is only with "m" that we have an acquaintance, the existence of *a* resting merely on an inference. Heretofore I have often appealed to this fact, in order to show that the evidence both of ponderable
and imponderable matter is of the same kind precisely: the existence of
properties which can only be accounted for by inferring the existence of
an appropriate matter to which these properties appertain.

So far, Hare seemed to be making a case for identifying the world
with the perception of it—the same subjectivist viewpoint that was
currently being propounded in Germany and which Tyler had
campaigned so vigorously against. But in the very next sentence he
drew back from this conclusion to affirm the reality of the matter
being inferred:

Yet I cannot concur in the idea that because it is only with "m" that we
are acquainted, the existence of *a* must not be inferred; so that bodies are
to be considered as constituted of their materialized powers.[23]

While Hare's argument is on the surface analogous to the Scottish
philosophers' contention that man cannot reasonably deny the
existence of that which presents itself to the senses, the consequences of Hare's variation of the argument, because of the different
kind of object to which it was applied, were radically different. He
had, in fact, drifted completely out of Scottish philosophy by pursuing it to its logical limit. For the argument of Reid and Stewart,
when applied to the subject of the imponderables, led to a conclusion that the celebrated common sense had to reject—there was a
"matter" possessing neither weight nor extension nor impenetrability. Closer adherents to the Scottish philosophy, like John Ware,
whose chief work was an enlarged edition of Smellie's *Natural History,* could never accept such an insane world as one in which there
was matter without weight. The "imponderables," he had protested
as early as 1819, were only properties of matter, not types of matter
in themselves.[24] And by "matter" he meant that which possessed
those properties commonsensically associated with matter—like extension, impenetrability, and a determinate weight.

The example of Robert Hare demonstrates how easily the Scottish

philosophy could be transformed into a rudimentary kind of operationism when it had to face questions that had never been entertained by Reid and Stewart. The belief that "that which has
properties is matter" is a judgment of the common sense, but the
common sense also dictated that there were certain specific properties always associated with matter. The undeniable existence of
those hypothetically inferred entities, the imponderables, forced a
judgment between these two dictates of common sense, for they
became incompatible with each other when the question of the imponderables came into the picture. To Hare, along with many
others, the first seemed more important. There were operations by
means of which one could achieve predictable mechanical effects
from any of the imponderable agents; this was enough to convince
Hare of their reality. Beginning with this conception of matter as an
operationally defined concept, Hare thought he could use rules of
logic to deduce a clear picture of the fundamental forces of the
universe and of the matter behind those forces. The most important
rule was the formal logical principle of non-contradiction:

The existence of repulsion and attraction being proved, it must be admitted that they are properties of matter; since the existence of a property, independently of matter is inconceivable. But being of a nature to
counteract each other, the repellent and attractive powers cannot coexist
in particles of the same kinds. There must, therefore, be a matter endowed with repulsion, distinct from that which is endowed with
attraction.[25]

The phenomena of chemistry, Hare thought, demonstrated that
there existed at least three different kinds of matter having properties that were incompatible with each other—an attractive matter,
a repulsive matter, and a matter having a variable, limited, and
contingent attraction.

Less imaginative scientists than Hare displayed a measure of
impatience with his kind of reasoning. Denison Olmsted, in his textbook of 1844, was unwilling to assign the name "matter" to anything
which did not have the two essential properties of extension and
impenetrability.[26] Earlier, he had misunderstood Hare's rather
subtle distinction between matter and properties of matter, upon
which his willingness to admit a repulsive matter depended. Reviewing Hare's argument against Davy's hypothesis that heat was motion
among particles, he accused Hare of entertaining the outmoded

belief that powers resided outside the bodies to which they belonged. Since the time of Newton, he said, philosophers had generally held that they knew nothing of the cause of gravitation. Certainly no one had suggested, as he thought Hare was arguing, that
there was matter in which attraction resided, "distinct from the
bodies themselves, which exert this influence on each other." To
Olmsted, this sounded reminiscent of the "elements and vortices" of
Descartes, or the "subtile ether" which, he carefully said, Newton
had barely allowed a place "in the form of a query in a corner of the
appendix" to his *Optics*.[27]

Hare, of course, had made no such claim; but the misunderstanding was probably honest, and it reveals the radically different viewpoints of the two men. Olmsted, looking at the concept of matter
from the viewpoint of the physicist, thought in terms of *bodies*,
which he defined as a collection of matter existing in a separate
form; and when he spoke of "particles" he meant "*the smallest parts
into which a body may be supposed to be divided by mechanical
means*." Inquiries respecting the composition of bodies he was willing to leave to chemistry.[28]

But it was exactly those questions that Olmsted would leave to
chemistry that interested Hare; when he spoke of an attractive
matter he was not thinking of a matter existing distinct from the
bodies that displayed attraction, but he was thinking of the composition of the bodies themselves.[29] His argument was simply that
repulsive matter and attractive matter were the fundamental chemical units of the universe, and all bodies contained both in varying
proportions.

The explanatory possibilities of Hare's "repulsive matter" were
immense, even though it was painfully obvious that such matter
must act contrary to the Newtonian law. Hare, in a published
argument with William Whewell, even demonstrated that the inverse square law could just as well be derived from the postulate of
repulsion as it could from attraction. It therefore made no real
difference what one assumed about matter.[30] As Lewis C. Beck
showed, one could also use the concept to explain all the phenomena
of heat. Explaining calorific repulsion to his students, he noted, "that
its particles repel one another, is proved by observing that it flies off
from a heated body"—thus heat radiation could be explained by a
visual image—"and that it is attracted by other substances, is

equally manifest from the tendency it has to penetrate their particles and be retained by them"—this explained heat conduction and the otherwise puzzling observation that heat always flowed from a hot body to a colder one until an equilibrium of temperature was reached.[31] Beck explained that he had adopted the view that caloric was a distinct substance because it "affords a more easy explanation of most of the chemical phenomena," and his textbook demonstrated how all the other phenomena of heat could be explained by the hypothesis.[32]

Hare and Beck went further into speculations about the nature of matter than most men of their time; but even they refrained from attempting to construct a world from their postulates. Both of them characteristically failed to display an interest in speculations that seemed to have no immediate bearing on their theoretical needs. Their interest was, as Beck said, to find a "more easy explanation" for the phenomena they were studying. For example, it was not necessary, for their explanatory needs, to settle questions like that of the infinite divisibility of matter. "The atomic theory," Beck observed,

proceeds on the supposition that every body is an assemblage of minute, solid particles, without considering whether a further division of them be possible or not. The question of the infinite divisibility of matter is not therefore involved; the theory only assumes that matter is not, in fact, infinitely divided.[33]

Beck even admired Ampere's willingness to assume as fundamental the observed attraction and repulsion between electrical wires, instead of trying to explain these facts. His point, demonstrating the residue of Scottish philosophy that he had retained, was that there was no real need to explain observed facts; one rather accepted them and then used them to explain that which was not so obvious.[34]

Lardner Vanuxem, however, was bolder than Beck and Hare, and less willing to stop at the point where his immediate research demands ended. In a "preliminary" essay of 1827, he tried to prove that the ultimate principles of chemistry, natural philosophy, and physiology could be deduced from the existence of the two kinds of matter—self-repellent and self-attractive. The union, in varying proportions, of these two types of matter explained all the phenomena of the physical world. The laws which governed the one type of matter were the converse of those which governed the other.[35]

Vanuxem's notion could indeed "explain" everything—for the very reason that in actuality it was entirely devoid of content. Consider, for example, his interpretation of the now classic experiments of Dulong and Petit, the two French chemists who had made a number of experiments with metals in order to determine their specific heats. Their results had yielded the Baconian fact that specific heat multiplied by the atomic weight of a substance was a constant. They, therefore, concluded that the atoms of all simple bodies had exactly the same heat capacity. But, according to Vanuxem, their deduction was in error; what the experiments had actually shown was *"that the capacity of atoms for heat increases in the ratio of their levity,* and decreases in the ratio of their density."[36] The true result of their experiments was, therefore, to confirm the action of the two classes upon each other. This should make obvious the reason for Vanuxem's success in explaining all of nature with his principle. He had strayed so far from contemporary Baconianism that there was absolutely *no* data about the world at the basis of his reasoning; instead, it was founded on an empty tautology which could serve no conceivable utility; namely, that as weight (heaviness) increases, levity (lightness) decreases. He could then interpret the experiments of Dulong and Petit as confirmatory of his hypothesis, because to his mind they were multiplying the atom's "lightness" (caloric) by its "heaviness" (absence of caloric). If there had been any way to do such a mathematical operation, it might indeed have produced a constant; but how it could have given any information about the world was a matter that Vanuxem did not succeed in communicating.

Vanuxem was so possessed with the idea of his two classes of matter that he used it to construct a scale of organization for the entire animal and plant world. All organized beings were arranged on an ascending scale from the zoophytes, which contained the most "common matter," to the plants, which contained the most "ethereal matter." What this scale amounted to, briefly, was an arrangement of animals and plants in terms of their inflammability—zoophytes would not burn and were at the bottom of the scale, plants burned readily and were therefore at the top of the scale. Man, along with the rest of the mammalia, was in the center—he would burn a little more readily than reptiles and not quite so well as birds.

It is easy to see, at this point, that Vanuxem had been wholly captured by his two types of matter. He tried to go even further with his mathematical considerations on the ratio of mixture. He noted, for example, that plants were without locomotion, which seemed to him to be the case with all organized beings formed either of an excess of ethereal matter, or of an excess of common matter, as the zoophytes. This led him to establish still another law. There were analogies between beings in equivalent positions in the bottom half and the top half of the scale. The greatest analogies were between plants and zoophytes, the woody parts of one corresponding to the stony parts of the other, the sexual parts of plants corresponding to the animal parts of the zoophytes. There were "other analogies" (unspecified) between insects and crustacea, birds and fish. Presumably, although he seems to have missed the implication, man, being in the exact center of the scale, was unique in being analogous only to himself.[37]

Speculations like these could hardly be justified in terms of the received opinions about the nature of scientific method. Although Vanuxem certainly carried the chemical ideas to their extreme manifestation, he could still not be dismissed as belonging to a lunatic fringe. He was a well-educated scientist—an 1819 graduate of the Paris School of Mines—and for the seven years before writing this essay he had served as professor of chemistry and mineralogy at South Carolina College. His lasting scientific reputation, however, rested upon his geological publications, especially his Geology of New York, Third District (1842); and to him also belongs the honor of having been the first American to write a paper in stratigraphic geology making primary use of fossils.[38]

Even though Vanuxem could not simply be written off as a lunatic, it could still be said by those who wanted to dismiss or excuse his speculations that he was not primarily a chemist, and had therefore been writing out of his field in the Essay.[39] This charge could not have been made before a relatively advanced degree of specialization came to be the norm, but by Vanuxem's time the charge carried some weight. But even if Vanuxem could be explained away by the cautious, what was one to do with men like

Hare and Beck? They were both professional chemists; they were, in fact, probably the two most respected chemists in America in their own time. It was particularly unsettling to note that they had even dropped the notion that a scientific theory had to be judged either true or false before it could be judged for its utility—an idea that is not so strongly insisted upon in modern physics, but which was generally believed then.[40] It was, in fact, a logical result of the idea that a theory was no more than the generalization of a fact. Hare and Beck, however, never spoke in terms of the truth or falsity of a theory, but only of its *utility*—a concept which they, unlike some others of the period, did not confuse with truth. Their viewpoint seemed to be that theories must group phenomena in order to make them workable, and therefore any theory which successfully handled all the relevant phenomena was a good theory—an idea that Joseph Henry later made explicit and which has come to be widely held.[41] Furthermore, both Hare and Beck knew that in many cases rival theories could serve the purpose equally well, and that there might be no satisfactory way of choosing between them. Hare, as observed before, had demonstrated that the inverse square law would hold whether the fundamental relations between particles were of attraction or of repulsion; consequently it did not seem necessary to him to decide that the relations were either attractive or repulsive, in any exclusive sense. Whewell, taking a more conservative approach and still clinging to the criterion of universality,[42] had urged that if it were not true that all matter had weight then the very idea of weight was useless.

Even in view of the excesses of Vanuxem, and Hare's radical redefinition of matter, Beck is still probably the best example of a scientist who quietly dropped his Baconianism without even acknowledging that he was transgressing the rules. Beck was a follower of Franklin in his electrical theory—all bodies were assumed to have a "normal" amount of a single electrical fluid; positive electricity was manifested by bodies having an excess of the fluid, while negative electricity was a result of a deficiency in the fluid. He noted that there was one significant defect in the original theory; the repulsion observed between two positively electrified bodies was easily accounted for, but that between negatively charged bodies was left unexplained. This could not be referred to repulsion between particles of the electric fluid, for the bodies were supposed to contain less than their normal amount, and it was inconceivable that a mere deficiency could produce an effect which counteracted the normal action. This was the one great problem confronting the one-fluid theorist, but Beck thought it could be solved quite easily. To the theory one had only to add the additional hypothesis that particles of matter uncombined with electricity exert a repulsive action upon one another. Thus, repulsion of two negatively electrified bodies was considered the result of repulsion of particles of matter.[43] In one bold stroke Beck had ignored the Newtonian canon against multiplying hypotheses, and he had declared, for the sake of the theory, that on a fundamental level there obtained an opposite condition from that demanded by the law of universal gravitation. Furthermore, the additional concept that Beck called to the defense of his theory was in principle untestable, for, by the hypothesis, no matter existed anywhere in nature which was uncombined with electricity. The sole function of the hypothesis was to save the one-fluid theory of electricity, but this caused Beck no evident embarrassment.

The necessary deduction that Newton's laws were not really fundamental was enough to make the more conventional Denison Olmsted posit two fluids, even though he knew that in so doing he was violating another Newtonian canon—employing two agents to effect a given purpose, when a single one would have been competent to its production.[44] Even the most avowed Baconians among the natural philosophers were reduced to choosing the lesser of two evils.

But Beck's greatest contribution toward confounding Baconianism by embarrassing its fundamental postulates was in making Ampere's work conform to the one-fluid theory. Ampere, working on the assumption of the rival two-fluid theory of electricity—that there exist two distinct material fluids, the positive and negative (or vitreous and resinous)—had done pioneer work in demonstrating and reducing to a mathematical basis the connection between magnetism and electricity. His success, incidentally, had generally been thought to have confirmed the two-fluid theory. Beck, however, refused to be converted, for he discovered that by changing a few terms he could use all of Ampere's formulae and arrive at the same conclusions on the assumptions of his one-fluid theory. It really seemed that it made no difference whether electricity was composed

THE LIMITS OF BACONIANISM

of one or two fluids, or as later work was to show, of no fluids at all. In other words, the "facts" at this level, far from being the touchstone of theories, did not even seem to be a relevant issue.[45] Even Olmsted had to grant that nearly every electrical phenomenon could be perfectly explained in accordance with either hypothesis, and that no *experimentum crucis had yet* been found, although it never occurred to him that one *might not* be found.[46]

The deep-seated association between religious attitudes and scientific thought which was brought out earlier is admirably illustrated by the main subjects of this chapter so far—Hare, Beck, Vanuxem, and Olmsted. Olmsted, who has been used as a kind of Baconian foil to display the nonconformity of the others, had spent two years studying theology at Yale, at the time (1815–1817) a citadel of conventional religion. The other three were decidedly unconventional, although all spent their last years in recondite religious studies. Vanuxem retired to a farm to spend most of the last fifteen years of his life studying Eastern religions, phrenology, and Egyptology; the fruit of Beck's last years was a large number of religious manuscripts, including a work entitled "Claims of Religion upon Science." [47] But Robert Hare is the best example of a man's scientific views leading him directly into a radical religious mysticism. He had been a skeptic most of his life, and, if his friend Silliman is to be believed, a misplaced man of the Enlightenment in his religious beliefs.[48] But during the great wave of spiritualism during the 1850s Hare became a convert and leading propagandist for the movement. He reported that he had investigated the claims of spiritualism scientifically and found that he had no choice but to give his assent. How close his speculations on the nature of matter had already brought him to spiritualism is well-illustrated in the passage where he explains the reason for his conversion:

Confining the range of my philosophy to the laws of motion, magnificently illustrated by the innumerable solar systems, but no less operative in every minute mechanical movement, I hold that I could only come to the same conclusion as Faraday, *that if tables when associated with human beings moved,* it must in some way be due to those beings, since, agreeably to all experience of the laws of matter in the material world, inanimate bodies cannot originate motion. But as when the planetary motions are considered, any hypothesis fails which does not account for the rationality of the result, and therefore involves the agency not only of a powerful but of a rational cause; so the manifestations of Spiritualism, involving both reason and power, might consistently justify me in looking for agents endowed with the reason and power manifested by the phenomena. This power being *invisible* and *imponderable*, and at the same time *rational*, there was no alternative but to consider it as spiritual, no less than that to which the planetary motion is due.[49]

It was not very far from a weightless matter acting contrary to the fundamental laws of matter to a spiritual "matter" possessing reason; both were admitted by Hare on the same operational grounds—the existence of phenomena which he thought could not be explained without the assumption. Although the attitudes toward the nature of theory held by Hare, Beck, and Vanuxem were a great deal more sophisticated and were ultimately to prove a great deal more useful than those of Olmsted, it must still be noted that, in the absence of further refinements, the same kind of operationism which admitted the imponderable substances could quite legitimately admit a spiritual world.

Under the heading of changed research interests, my concern has been almost entirely with chemistry, principally because this subject is less well-known and it requires more explanation to make obvious the changing viewpoint demanded by developments in that science. But it was by no means the only science that found itself in a troubled state by the end of the first quarter of the nineteenth century. Quite the contrary, the research interests in a great number of sciences began to change at the same time. In each case, the questions being asked were simply unanswerable in terms of the received conception as to correct method. In historical geology, for example, one could not observe a succession of mountains being formed and then reach some automatic generalization about the formation of mountains. Instead, geologists had to observe a great number of mountains in various stages of formation when they had no thoroughly empirical way of even knowing which stages were earlier and which later—or, indeed, if all were not simultaneous creations. Paleontologists, attempting to classify strata in terms of their fossil contents, were told that the significance of their fossils could never be known until the true order of superposition of the rocks was established independently, the very aim the American paleontologist T. A. Conrad and others hoped to achieve by means of the fossils.[50] One had to begin, it seemed, with a "hypothesis" either about the rocks or about the fossils. There seemed to be no beginning point entirely free of the hated hypothesis. The geologists and paleontologists, along with the anthropologists, the plant morphologists,[51] and the systematic zoologists, were having fully as much difficulty in viewing their portion of "the arcana of nature" directly as were the chemists and physicists. The interests, in each case, had shifted from the collection of data and the explanation of observable individual facts to questions of origins, of relationships, and of changes which by no conceivable means could be seen in action.

Recognition of the growing disparity between the accumulation of facts and the explanation of those facts was as evident in geology as it was in chemistry, and this recognition resulted in the same kind of departure from conventional Baconianism. Edward Hitchcock had occasion to review two geological works during the 1820s, one in 1824 and the other in 1829. A comparison of his changing opinions in these two reviews is instructive. In the earlier article he had noted the rapid accumulation of geological knowledge and remarked that it was hardly safe for one to announce a new geological fact if he would not stand corrected in the next issue of the journals. Nevertheless, his endorsement of the work under review was on the grounds that the authors' classifications did not imply hypotheses and that the work was free from "doubtful theoretical views." He was able to endorse the authors' statement that some rocks appear to be of igneous origin only with an apologetic aside. He concluded that,

We are indifferent whether we are called Neptunians or Vulcanists; and also, whether we have any general theory on the subject, or only some points of apparently contradictory theories. Just so far as undeniable facts lead us, we wish to follow, but no farther. We do not believe the time has yet arrived, in which it is possible to make any very extensive, correct generalizations in regard to the original formation of rocks.[52]

The second article, written five years later, struck quite a different tone. The avoidance of theory, which he had praised in the previous article, had by then become "unreasonable skepticism," and while Hitchcock admitted that he had long been among the number of confirmed skeptics of all geological hypothesis, he had "become wearied with hovering so long over what seemed to us the shoreless ocean of uncertainty and conjecture." The outcome of his reflections was an unqualified acceptance of Cordier's theory of a central heat, which, Hitchcock admitted, rested primarily on an analogy between the increase of temperature in mine shafts and the increase in temperature to the earth's center.[53]

Fundamentally, the problem of the scientists discussed in this chapter was one of trying to cling to a set of methodological rules and assumptions about the world, when it was becominng clear that, in order to avoid breaking one rule, they would have to break another. As conflicts and difficulties arose, they turned—as scientists generally do in a time of crisis—to ad hoc modifications of theories in an effort to eliminate the developing conflicts and pursue their investigations.[54] Beck, Vanuxen, and Hare had ignored one after another of the canons of philosophizing and had gone beyond the fundamental laws of matter as the occasion demanded. All three had abandoned the notion that the fundamental relation between particles was one of attraction; Beck and Hare had shown that rival theories would do quite as well for explaining phenomena and for calculation; Beck had "multiplied hypotheses" in order to reconcile fact with theory and had postulated at least one hypothesis which was in principle untestable. Even Olmsted, in order to save the fundamental nature of Newton's laws, had found it necessary to make a flagrant violation of the canon of simplicity. Hitchcock, becoming wearied of having no explanation for his facts, had reversed his position almost to the extent of becoming a uniformitarian. No matter how valiantly he tried, there seemed to be no complete escape for anyone working in these areas.

Hitchcock, Hare, and all the other men discussed in this chapter were driven to violate their methodological principles because of their desire to explain the phenomena they were studying and their dissatisfaction with the apparently sterile accumulation of facts. It is true that a part of the difficulty lay in their unwillingness to accept the idea that reference to a general law was the only type of explanation that science could give them. This was a tautology, Olmstead had said, and dismissed the idea to turn to an analogical grouping of chemical facts as the only path open to chemists. No pyramid of laws, as Mark Hopkins observed, could ever remove the mystery; each higher law merely moved it one step further away.[55] Olmsted and Hopkins were demanding something science could not give them—explanation in terms of purpose—and while the concept of purpose could still be made to work in biology without flagrant violation of Baconian canons of method, it was becoming increasingly unsatisfactory as a general interpretation of nature. And as the Baconian philosophy of science became more and more unsatisfactory, so too did the idea of purpose in nature.

༄

CHAPTER VII

Finalism, Positivism, and

Scientific Explanation

Maintaining the scientific orthodoxy developed in America became an increasingly difficult task in the face of growing knowledge and the consequent growing specialization of interests. Scientists were forced to impose severe limitations upon themselves concerning both method and objects of study. They had to forbid such things as chemical speculations about physiological processes, inquiries into geological origins, and any effort to explain the "ultimate forces" of nature—electricity, magnetism, heat, and other such entities. All of these things involved matters beyond the reach of the senses and were consequently forbidden by the peculiar form of Baconianism developed in nineteenth-century America.

Increasingly confronted with the inadequacy of Baconian philosophy for scientific explanation, a large and vocal group of thinkers sought to accommodate these deficiencies by denying to science in general any explanatory function. Scientific investigation, they maintained, was to be restricted entirely to collection, description, and classification. Anything further than this was said to be forever outside the domain of science. Growing quite naturally out of Baconian empiricism, this negative assertion was also in part a reaction against the sometimes excessively speculative predilection of the previous generation. There was no denying that the Enlightenment thinkers had often carried their theorizing "beyond the bounds of sound philosophy or common sense," as a then current expression would have it. The last part of the eighteenth century had been a period of particularly wild and irresponsible generalization in medicine and in the life sciences. It had been a time when it seemed natural to compliment an important medical system-builder with the statement that "to avoid the contagion of opinion, he had read no medical book for five whole years." [1]

Medical theorists had gone from extreme to extreme in constructing their systems. One group would begin its work by constructing a truly overwhelming nosology. Every supposed variation of a disease and every imaginable combination of symptoms would be named a distinct species, and each would be said to require its own elaborate treatment. Pulmonary tuberculosis, for example, by many theorists was broken down into twenty different species. Reacting to these excesses, and the consequent impossibility of practicing within such a framework, other theorists proclaimed that there was but a single disease, manifesting itself in different "states." Each of these theorists, of course, had his own ideas concerning the seat of disease, its cause, and the single treatment appropriate for every manifestation of illness.[2]

Under such conditions as these, theoretical doctrines had risen and fallen in such rhythmic succession that a sober onlooker could not be blamed if he came to look askance at theory per se. Early in the nineteenth century, Thomas Jefferson, himself quite facile at theorizing in other areas, expressed a common attitude toward some of these doctrinal excesses:

the adventurous physician goes on, and substitutes presumption for knowledge. . . . He establishes for his guide some fanciful theory of corpuscular attraction, of chemical agency, of mechanical powers, of stimuli, of irritability accumulated or exhausted, of depletion by the lancet or repletion by mercury, or some other ingenious dream, which lets him into nature's secrets at short hand. On the principles which he thus assumes, he forms his table of nosology, arrays his diseases into families and extends his curative treatment, by analogy, to all the cases which he has thus arbitrarily marshaled together. I have lived to see the disciples of Hoffman, Boerhaave, Stahl, Cullen, Brown, succeed one another like the shifting figures of a magic lantern.[3]

On another occasion, in a letter to Rush, Jefferson related the story of Doctor Seymour, a (perhaps fictitious) Virginia physician who tried to employ his medical skill to cure the ailments of tobacco plants. "He bled some, administered lotions to others, sprinkled powders on a third class, and so on—they only withered and perished the faster." [4]

The lesson that was in all of this for Jefferson—as for many others of the period—was that the physician should abandon "hypothesis for sober facts." He should set the highest value on clinical observation, the lowest on "visionary theories." [5] One finds such complaints and such recommendations for improvement in all quarters during the first few decades of the nineteenth century. "System has so often fallen under the stroke of experience," observed James Maclurg, "that we have been led to imagine a natural enmity between them." [6] Indeed, many did imagine an opposition between system and experience, and they generally decided unequivocally in favor of experience. It was not only the amateur like Jefferson who protested against medical theorizing. Many professional scientists as well were deeply disturbed by the spectacle their colleagues had presented to the public, and were determined to call a halt to their excesses. J. Augustine Smith, in his 1828 address before the New York Medical Association, gave the following "recipe" for a medical theory:

Take a fact true in one case at least, if it will hold in half a dozen, so much the better. Extend it to all times, places, seasons, persons, and circumstances. Let it embrace every malady that flesh is heir to; let it comprehend the whole human family, including men, women, and children, of every age, sex, colour, condition, and profession, and you have a theory possessing all the usefulness, truth, and stability which so very wise a proceeding can be supposed to confer.

Smith painted a bleak picture of the state of medical thought in his own time. It was confused, shallow, imitative, or downright murderous, he thought; composed of "assertions which are apocryphal and systems which are false, the whole rendered more humiliating and disgusting by a copious admixture of bad feelings and angry passions." But he also understood the source of the difficulty and he knew what had to be done. "We must have gone astray," he said,

we must have failed to observe those dictates of right reason which, as rules of philosophizing, have conducted other inquirers, with serene tempers and kindly dispositions, to certainty and truth. Here then must be the fault; here must be the source of all our errors, our broils, and our contests. OUR PHILOSOPHY IS WRONG AND REQUIRES TO BE CORRECTED.

In other words, Smith declared that medical thinkers had somehow drifted away from the true philosophy—the Baconianism that had previously served them so well and was still serving inquirers with "serene tempers and kindly dispositions"—and they could solve all their difficulties by merely returning to it and reaffirming their allegiance. Furthermore, Smith thought, it was fruitless as well as foolish to waste one's time in a search for general laws. It was not that he denied the existence of such laws, but he was content to reserve their discovery for one of those superior geniuses who, "born at a lucky moment are destined to shed light and lustre upon the world." For the present, physicians should attempt to discover the natural history of individual diseases, deal with them one at a time, and control them and their effects so far as possible. This was a modest program, perhaps, but he thought it the best one available. "It is our business," Smith concluded, "to observe well, to observe long, and to observe all." [7]

Smith was representative of those who thought that the proper way to cure the ills of their Baconian philosophy was to put more Baconianism into it. From this viewpoint, the fault did not appear to be that the philosophy was inadequate or in need of refinement, but that it had not been consistently applied. Those who spoke in this manner typically based their case upon the claim that "reasoning" too often went "beyond the facts." This kind of attack upon reasoning, which lasted throughout the period, was particularly characteristic of the decade of the 1820s, for it was in that decade that large numbers of thoughtful men first began to realize that reasoning could lead to "unreasonable" conclusions.

Although its earliest manifestation was in the life sciences, the issue also became particularly acute in geology, another science which had been the subject of excessive romantic theorizing during the eighteenth century. One prominent school of geological thought would have the entire earth formed by the action of water; its chief rival postulated earth-shattering explosions and conjured up visions of gigantic fissures opening up to send forth streams of molten matter which formed the topographic features of the earth. Here again, a reaction was produced among the more cautious scientists, many of whom felt that such excesses were detrimental to geology's stand as a reliable, empirical science. Benjamin Silliman, in the early part of the period, at least, was a spokesman for those who thought that much of the geological work then being done was grossly misdirected. "Geology, at the present day," he affirmed,

Means not a merely theoretical and usually a visionary and baseless speculation, concerning the origin of the globe, but, on the contrary, the *result of actual examination into the nature, structure, and arrangement of the materials of which it is composed.*[8]

A close study of Silliman's statement provides a number of revealing insights into the attitudes and typical mental associations of that period. "Theoretical" is prefixed by "merely," and the "merely theoretical" is set against "actual examination." Furthermore, the "merely theoretical" is usually "visionary and baseless"—"theory" has almost become equated with "dream." Most importantly, this statement would completely deny the legitimacy of the historical dimension in geological studies, confining the geologist to the description of present forms. It was the historical aspect that seemed least amenable to Baconian philosophy, and it was accordingly the part of geology subjected to the severest criticisms. "Where wast thou when I laid the foundations of the earth," the cautious would have God ask the presumptuous geologist who meddled into matters beyond his power of perception.[9]

Amos Eaton, in 1818, published a largely uninspired and conventional *Index to the Geology of the Northern States,* but he made the unfortunate error of attaching an Appendix which he called "Reflections on the History and Structure of the Earth." A sample of the reactions to it is instructive. C. S. Rafinesque, in the *American Monthly Magazine and Critical Review,* denied that Eaton's Appendix should even be called "geology," for "geology describes the earth as it is," he said following Silliman in his denial of the historical dimension, ". . . and no one will venture to deny its conclusions, since they arise from facts and existing causes." The best that he could say for speculations like those of Eaton, however, was that they were "ingenious dreams" based upon nothing more solid than "suppositions, conjectures, fictions, presuppositions, probabilities, and plausible causes"—none of which had any legitimate place in Baconian science.[10] Eaton, good Baconian that he usually was, had clearly gone astray on this matter.

The rejection by the reviewer in the *North American Review* was a great deal milder—but still a rejection. "We have had speculative geologists enough," he thought, and it was now time to reason from facts. The speculations at the close of the pamphlet he did not think "its wisest or most valuable part."[11] Neither of the above writers confined himself to rejecting Eaton's speculation in particular—in fact, neither of them even appeared to consider it worthwhile to spend the time in making a good argument against it. It was speculation per se that they considered objectionable, and having shown that Eaton's "Reflections" were really "speculations," it seemed to them that their critical task had been accomplished.

Even the quite orthodox Horace Hayden received a mild reprimand from Silliman. In his *Geological Essays* (1820) Hayden attempted to account for the alluvial deposits along the Atlantic Coast of the United States. The chief agency in their formation, he decided, was a great "general current" which had flowed from the northeast over the entire continent at some unknown period. The proof that there had been such a current was the accumulation of sand, gravel, rolled pebbles, and boulders which were scattered about the region.[12] Hayden, being a religious man and a Biblical scholar, naturally tried to identify his current with the Noachian deluge. Silliman had no objection to this particular speculation, but he made it clear that Hayden had greatly exceeded the bounds of philosophizing. It would have been better, he said, if Hayden had excluded the "extraneous matter," and had contented himself with giving a "digested arrangement" of all the important facts connected with alluvium. Had Hayden done this, whatever theory was warranted would "flow naturally as an induction, according to the strict Baconian mode of philosophizing."[13] Ironically enough, a few years later when Silliman allowed his zeal to demonstrate the compatibility of science and religion to overcome his caution in the same manner that Hayden had, he was himself attacked by a reviewer in much the same terms.[14]

To the more conventional of the nineteenth-century scientists, it seemed that every new discovery, each scientific advance, had been magnified beyond all reasonable proportion and that the most extravagant claims had been made for it. Men were not content with merely observing and describing a new discovery and placing it in its proper position on the Baconian ladder of science. It seemed that they had to attempt a revision of the whole structure of science in order to show that the new discovery was at the foundation of it. Yale professor Denison Olmsted described a typical attitude toward the results of Franklin's important contribution to knowledge about electricity when he complained that after the identity of electricity with lightning had been proved, electrical theorists fancied that they had discovered the clue which would conduct them safely through the "labyrinth of nature." Everything which had not been satisfactorily accounted for was ascribed by the overzealous theorists to electricity. "They saw in it, not only the cause of thunder storms, but of storms in general," he protested,

of rain, snow, and hail; of whirlwinds and water spouts; of meteors and the aurora borealis; and finally of tides and comets and the motions of the heavenly bodies. Later electricians have found in the same agent the main spring of animal and vegetable life, and the grand catholicon which cures all diseases. Recent attempts have been made to establish the very identity of galvanic electricity and the nervous influence, by which the most important functions of animal life are controlled.[15]

Olmsted's selection of an example from chemistry was not at all fortuitous, for as has been suggested before, chemistry was quite often the culprit discipline when the Baconian rules were broken. And it was chemistry that was most often found in conflict with the leading ideological commitments of the times.[16]

One of the most important of the ideological commitments of that generation was a purposive view of nature—a belief that every natural activity subserved some specific purpose in God's plan for the world. The physical universe was an ingeniously contrived mechanism designed explicitly for the use of man and the glory of God.[17] Biological organisms were endowed by that same Creator with a force unique to life, a "vital force" which was the designated agent for achieving that organism's purpose. Beliefs such as these were cherished by the society at large, and the scientific profession, generally sharing the value structure of the society of which it was a part, was committed to their defense. The claim that a peculiar "life force" existed was, in fact, completely compatible with the mechanistic viewpoint of the time, for mechanical philosophy was fundamentally Newtonian; that is to say, positivistic.[18] Newton had foresworn any effort, at least in his published scientific writings, to attain to any explanations beyond the phenomena in question itself; that is, he provided no explanation for the cause of gravity, but merely generalized it to apply to all bodies.

While the poets may have rebelled against the implications of a mechanistic universe, by the beginning of the nineteenth century, at least, vitalists in science were generally unperturbed. They had so thoroughly assimilated mechanism to their own viewpoint that to challenge its descriptions seemed almost to be challenging the wisdom of God, who had by some mysterious process contrived the mechanism and was directing it toward His own inscrutable ends. It was well-known that a description did not amount to an explanation, and the mechanists claimed to do no more than describe how God directed the matter which He had made in its current form. The theologians could explain and the mechanists could describe; therefore there was no reason for one to interfere with the domain of the other, and a feeling of mutual dependence and aid could develop. This was the working arrangement arrived at by scientists and theologians, and majority opinion in both groups believed that their professional interests demanded the preservation of the arrangement.

This happy liaison, however, was severely threatened by the appearance of chemistry as a rival of mechanism, for the vitalists had a great deal more to fear from a chemical universe than they had from a mechanical one. The chemists were thought to be overreaching the bounds of philosophy in their attempt to go beyond a simple mechanical interpretation *with explanations that were not teleological,* and their announced aim was eventually to reach all the way to the fundamental forces of the universe and to the last minute particle of matter with a purely natural explanatory chain. Indeed, they were even discussing the creation of matter itself. No matter how cold and uninviting a universe mechanistic principles might in retrospect seem to suggest, those professing allegiance to them characteristically saved themselves from opprobrium—and perhaps from personal disenchantment as well—by a staunch avowal that their science dealt only with the outwardness of phenomena. The inner nature of the phenomena, and the cause of their behavior, were matters that were said to be hidden forever beyond the reach of their methods. Most important of all, while the mechanical philosophers dealt with matter as God had created it, chemical philosophers impertinently asked *how* He had created it and in arriving at answers, threatened to discard divine causation entirely. They seemed, in fact, to be playing the role of God Himself in their destruction of the given nature of things; and in their refusal to accept as fundamental the building blocks of the mechanists.[19]

Even though this heresy was recognized as being implicit in the chemical philosophy of the early nineteenth century, no one was willing to admit that *he* held such a position. In searching through the scientific literature of that period, nothing is more evident than the frequently reiterated denunciations of those who tried to explain too much by chemical principles. Yet, search as one may, he is hard put to find an offender. Nor were those who denounced the practice able to name offenders, other than "French materialism" in general, or some long-dead philosopher who had no acknowledged nineteenth-century followers. As so often happens in human history, the threat existed only as a dreaded possibility, which long anticipated its appearance as actuality. The fact is that at least until the third decade of the century, long after the defenses were erected, even the chemists generally held that science knew nothing of causes. At the very least they drew sharply defined limits to the range of applicability of their principles. Even Samuel L. Mitchill, who distinguished "modern" medical reasoning from "ancient" on grounds that the modern mode was chemical, insisted that despite chemistry's frequent efficacy in explaining animal functions,

we are not to imagine everything in life susceptible of chemical interpretation. What it is that enables the atoms composing a muscle to cohere, and the muscles to contract and perform great exertions of strength, we know not; but this we know very well, that we can never form a muscle by synthesis, or the putting together, in any artificial form, those substances which appear, from analysis, to constitute a muscle. There is something in animated existence, which eludes our most active researches, and which defies submission to either *mechanical* or *chemical* laws.[20]

According to the current viewpoint, scientific knowledge was limited not only to the outwardness of phenomena; it was also limited to the succession of phenomena in time. It was considered "unphilosophical" to allege any power connecting the phenomena *as a part of the scientific explanation.* A law was defined simply as a description of the way the processes of nature worked (as established by direct observation), and was consequently not to be thought of as an explanation of the processes themselves.[21] This meant that science could be denied any "real" explanatory power, and those wishing such insights could be referred elsewhere. "Science only describes; it does not explain" was one of the favorite clichés of the day—a cliché that had powerful ideological support.

Thus, when the theologian Tayler Lewis argued that science was utterly incapable of answering the deepest and most perplexing questions which man faced, he was certainly not saying anything new.[22] The real innovator in the issue, as a matter of fact, was James Dwight Dana, who took exception with Lewis' denunciation.[23] Had Lewis written such a statement twenty years earlier, however, it is not likely that he would have received any opposition on this point, for he would have been repeating commonplaces.[24] Men would have been far more likely to have wondered why he bothered to expound such an obvious truth.

Why chemistry should have appeared as a much greater threat than mechanical philosophy is not completely explained by the above considerations, for even though chemistry delved deeper, those who wished could always have found a mystery behind its explanations. As the moral philosopher and educator Mark Hopkins observed:

Since a general law is only an abstract name for a uniform mode of operation, which name can have no efficiency, [it follows] that the power which operates according to the law, must be immediately exerted in producing every individual effect; and that if the law be mysterious, the particular facts, from an observation of which the law was inferred, must, truly and philosophically speaking, be equally so.[25]

Perhaps it was merely that mechanism had been around longer, and there had been time for thinkers to learn that there was no necessary opposition between the mechanical and the vital; while chemistry—in the modern guise which it had only assumed in the late eighteenth century—was a comparative newcomer. It had in the 1790s achieved an immediate success as an explanatory principle, and had been welcomed as a replacement for mechanism. By the 1820s, however, a great number of deeply concerned individuals began to see a necessity for reversing some of its successes.

The sharp contrast between the receptions given two medical dissertations illustrated the radical disenchantment with chemistry that characterized the early nineteenth century. In 1768 Benjamin Rush was granted a medical degree from Edinburgh University for a dissertation entitled *De Coctione Ciborum in Ventriculo*, an attempted chemical explanation of the process of digestion.[26] In 1822 John Patten Emmet was granted a medical degree from the University of New York for a dissertation entitled *An Essay on the*

Chemistry of Animated Matter, in which he argued that "the more these [vital functions] are examined the more one must be convinced, that the proximate cause of all organic formation is chemical affinity." [27]

Rush's work was well received both at Edinburgh and in America. The faculty of medicine at the University of Pennsylvania was so impressed with it that in the year after his graduation they unanimously elected him professor of chemistry with the right to examine candidates for medical degrees. For his work some fifty years later, Emmet received no unanimous appointment. As a matter of fact, it was three years before he found a job. In 1825 he was appointed by Thomas Jefferson to a professorship at the University of Virginia— and although he later wrote prolifically on electromagnetism and other relatively innocuous subjects, he apparently never wrote another word on "the chemistry of animated matter."

In spite of the fact that Doctor Emmet had followed the convention of speaking in terms of "proximate cause"—implicitly leaving room for an "absolute cause" or "first cause" behind the phenomena —the profession had been outraged. Henry W. Ducachet, writing in the *American Medical Recorder,* denounced the work as "materialism and infidelity." Even though Ducachet admitted that inaugural dissertations were traditionally immune from public criticism, he insisted that an exception had to be allowed in the case of Doctor Emmet. He had no choice but to protest, he said, when he saw an inaugural essay devoted to the support of principles which should be abhorred as "pestilent to the community," branded as destructive of the "high and holy sanctions of morality" for its arrogant condemnation of principles "which have ever been admired by the learned and good." Ducachet also thought that his duty to the community demanded that he take the University officials to task for accepting such a thesis in the first place. [28]

Besides general denunciations on obviously religious grounds, like that of Ducachet, two specific charges were most often involved in the attack on chemistry—or upon scientific explanation in general. First, it was said that principles of other disciplines, not applicable to the science in question, were too often being used. Second, critics insisted that the search for causes was illusory as well as vain— illusory because man's knowledge was limited to the succession of phenomena, and vain because a knowledge of causes would be of no use to the scientist even if it were attainable. The first was the issue of reduction,[29] the second that of causation; both of these have ever since been matters of philosophical debate, and on both, vitalism (or finalism) and positivistic mechanism could come into coalition against the common enemy.

In 1825 Nathan R. Smith voiced the general objection to reduction and at the same time revealed what was thought to be at stake.

. . . if we enter another field of science, the laws of which we have not investigated, and reason from an effect to a cause, we shall almost certainly fall into an error, for we unconsciously attribute the effect to a cause, with the operations of which we are familiar. Thus, if we combine two inorganic substances, and, after a time, observe that a remarkable change has taken place in their properties, we at once, and with propriety, attribute it to the influence of chemical affinities; for we know that such substances are exclusively under the control of physical properties. But if we convey certain substances into the stomach of an animal, a polypus for instance, and upon removing them after a considerable time, discover that they have become very much changed, having received new chemical and sensible properties, it is altogether unwarrantable to infer that these changes are wholly the effect of chemical influence; for, as the stomach of this and of every other animal is imbued with cerrtain vital properties, and as we are not yet fully acquainted with their laws or uniform habits of action we cannot say how much, for instance, is to be attributed to their influence.

With this much of Smith's statement anyone could agree, for in spite of its vitalistic overtones it is an unimpeachable statement of scientific caution—at least in terms of the knowledge of the time. But in the very next sentence he makes it clear that he regards this as an argument against the *principle* of applying chemical concepts to physiology: ". . . that similar effects in two distinct sciences, the cause being known in one, are referrible to the same cause, is, in principle, false." Smith's argument was directed specifically against what he termed the commonly held belief that digestion was accomplished by the chemical action of a gastric fluid. For this opinion he would substitute a simple mechanical explanation, which he thought more warranted by sound principles of scientific reasoning. All that men knew about digestion, Smith argued, was that when food came into contact with the walls of the stomach an important change immediately took place. If Smith understood the relation between cause and effect, he reasoned, it meant that one could not interpose a new link in the chain of events for the sake of aiding the imagination or for completing preconceived notions. One must, then, in the absence of further knowledge, conclude that digestion was effected by the physical action of the mucous membrane of the stomach.[30]

The digestion of food was to be accounted for on the basis of something in the body itself which was not operative outside the body. No matter how "mechanistic" it might appear, explanation in terms of the mechanical action of *living* solids was acceptable within the vitalistic frame of reference, for it left the cause of the action a great mystery, and it was obviously peculiar to a living body.

Because chemistry reached into the very nature of things, or so it was feared, while mechanics only described still unexplained behavior, chemical explanations were utterly unacceptable to a great number of people like Smith. Shortly after the publication of Smith's statement, another distinguished physician offered "some observations in objection to the commonly received doctrines of Digestion." It was probably the only time that a writer in Chapman's *Philadelphia Journal* ever acknowledged agreement with a writer in Eberle's *Review*, but D. Francis Condie cited Smith with obvious approval. Condie knew that the process of digestion had something to do with decomposition and recombination of food, but he did not think that one should even look for an explanation of the process in the laws of chemistry. "The vital power of the digestive organs," he said, "completely overpower the tendency to fermentation in the substances employed for food, while it induces in them a modification totally inexplicable by the laws of chemical affinity." [31] The strength of the language characteristically used by critics of chemical explanation is a rough indication of the tenacity of their views—and it also is a measure of their apprehension.

As Benjamin Rush had pointed out many years earlier, "to observe is to think, and to think, is to reason in medicine." This was why, he explained, one found theories in the work of any physician, "even of those who preface their works by declaiming against idle and visionary speculations." [32] In peremptorily reading chemistry out of physiology, the vitalists were themselves advancing an hypothesis, which was stated directly by Nathaniel Chapman: "No chemical operation of any sort can possibly take place in the living state, and all views predicated on it must be fallacious." Chapman's reasons for rejecting chemical explanations were the same as those of Condie—the function of the vital power, by its very definition, was to overcome chemical affinities and attractions. He confessed that in abandoning the chemical theory of animal heat he "left the subject in great obscurity, having no satisfactory explanation of the process to substitute." [33] This last sentence from Chapman gives the key to the success of the vitalistic attack upon chemical explanations, and it also provides another link with the positivistic mechanism with which it operated in coalition. For although those who protested against chemical explanations did indeed have a theory, it was quite easy to pass it off as no theory but as simple recognition of an empirical fact signaling a limit to the possibility of knowledge. A negative criticism, resting on the currently inexplicable, has historically been the main defense of vitalism, and even when they were theorizing outrageously, early nineteenth-century vitalists characteristically affirmed their pure empiricism.[34]

Chapman was one of nineteenth-century America's most vigorous campaigners against chemistry. It was largely because of his efforts that Robert Hare was forced to relinquish the right to examine candidates for the M.D. degree, a prerogative of the chair of chemistry since 1769, in order to gain the chair in the medical faculty of the University of Pennsylvania in 1818.[35] According to Thomas Cooper, Chapman actually thought the chair of chemistry ought to be separated from the medical department and all the requirements in chemistry dropped, for he considered chemical knowledge purely ornamental and would allow it no useful role in medical education.[36] In spite of his antipathy to chemistry, however, Chapman quite frequently used the appeal to ignorance as a deliberate technique in suggesting a connection between his own physiological ideas and ultimate physical agents. In expounding his doctrine of "sympathy"—a kind of occult transfer of powers from one part of the body to another, which he would substitute for chemical or neurological theories of the modus operandi of medicines—he explained that in using the term he meant only to denote, like chemical affinity, caloric, and other such expressions, a principle, or power which one would only know from the experience of its effects, "the precise essence or nature being occult, and concealed." [37]

The analogy Chapman made between "sympathy" and the ultimate physical agents was a well-chosen one for his purposes, for those who approached chemistry from a positivistic viewpoint were arguing that the chain of explanations had to end when one reached these agents. Olmsted, writing against one of Robert Hare's speculations, noted that since the time of Newton it had been the prevailing opinion among philosophers that one understood "attraction" merely as an undeniable *fact* of the physical world; one knew nothing of its cause, and was not likely to gain such knowledge. "In my view," Olmsted insisted.

our reasonings on physical subjects must stop when we arrive at one of those principles denominated *ultimate agents,* namely, attraction, heat, light, electricity, and magnetism; . . . all attempts to ascertain the nature of these agents, have hitherto proved unsuccessful; and . . . in the present state of our knowledge, we have no means of determining whether they severally depend on the operations of peculiar material fluids or not.[38]

Olmsted, as had Chapman, converted a deficiency of present knowledge into a permanently limiting condition for scientific reasoning itself.[39]

Considering the state of physiological knowledge at the time, there was some justification for Chapman's attitude; and there was also some justification for cautious use of a "vital principle" which maintained the organism and directed it toward some goal or purpose. It was a fact that the known laws of chemistry could not explain either life or the apparent purposefulness manifested by each organ. In pointing to areas of ignorance, the vitalists could find ample material, and ignorance did indeed seem to increase in fields that were most crucial for their argument. Mitchill, for example, had noted earlier that chemistry seemed much more successfully applied to those parts of the human body that had the "lowest degree of animation"—such as the teeth, bones, and fat.[40]

Chapman was particularly good at raising embarrassing questions for those who would explain life chemically. Why, for example, did not the animal body obey the law of heat flow if it was merely a chemical system? It was well-known that heat flowed from a warm to a colder body until an equilibrium of temperature was reached, he reflected, but animal bodies certainly did not obey this law; instead they managed to preserve an equable temperature in every exposure, and this, he thought, was enough to confirm the argument of vitalism.[41] In another instance he detailed an elaborate series of experiments which demonstrated that materials could not be taken directly into the blood. This seemed to refute the humoralist doctrine of the modus operandi of medicines as well as their doctrine of diseased fluids.[42]

Others made capital of the puzzling fact that when "morbific matter" was carried to different organs it produced different symptoms. This was difficult to square with the chemical hypothesis, because it seemed that the results ought to be uniform if the action were to be attributed to the introduced matter. The fact that the effects did vary, and that they invariably had an "analogy" with the "predominant vital forces" of the affected organ, was urged as a powerful argument against the thesis of purely chemical activity. This particular argument, which was the thesis of the 1830 Boylston Prize Dissertation, makes sense only within the vitalistic framework with its prior assumption that the body is not itself a chemical system, and consequently the organs of the body contribute no *chemical* factors to the reaction. The fallacy, however, was not noticed by the writer's contemporaries in the medical profession— probably because they had neither the inclination nor the logical tools to detect it.[43]

The vitalists, it would seem, had a good case so long as one spoke only in terms of existing knowledge—and the naive positivism of the day actually encouraged such a limitation. By the same token, there were quite convincing arguments in favor of the finalism which all vitalists shared. William Whewell, for example, observed that the function of every organ had been discovered by starting from an assumption that it must have some use; so far, investigators had not been disappointed. "The study of comparative anatomy," Whewell concluded, "is the study of the adaption of animal structures to their purposes." In the absence of a mechanism which could explain the fact of adaptation, the concept of a God sitting outside the process and directing it for predetermined purposes seemed the most reasonable explanation available. Anything else involved the attribution of a purpose to the process itself, a form of pantheism which early nineteenth-century thinkers were most anxious to avoid.[44]

John D. Godman was illustrative of the scientist who was seeking explanations, but who still could not withhold his assent to a moderate use of a "vital principle." Godman was one of the few Americans of his period who would uncompromisingly affirm the comprehensive lawfulness of the natural universe. At bottom was his conviction that "the animal economy exhibits the same harmony which is to be discovered in all parts of the great system and that in it nothing is effected, without the aid of the ordinary known laws of nature." [45] He disliked the "mystical jargon" in common use for describing the phenomena of life, and he held that the laws of chemical affinity and of mechanics were valid within the body as well as without it.

But while he vigorously denied that there was any suspension of universal laws by a vital principle, at the same time Godman had to concede that there was a fundamental difference in physical activity within the body, and he employed the term "Vital principle" to mean a modifying agent of some unknown sort that governed the process.[46] Godman also found it necessary to use the term "sympathy," but for him this did not imply, as with Chapman, a dogmatic rejection of further explanations. It was simply a fact that when one viewed the relations between the different parts of the system he could not help but note the dependence of the different organs and functions on each other. In the absence of exact knowledge, he thought that one should at least have a name for this condition, but he refused to consider such naming as the equivalent of an explanation.[47]

At the time he composed the above arguments, Godman was still, as an early biographer put it, "a French materialist." [48] Chapman, on the other hand, was a devout Christian, a personal friend of Lord Brougham and Dugald Stewart, and a firm adherent of the Scottish philosophy. Yet, even though Godman believed that physical explanations would someday be provided for "sympathy" and also for "vitality," he and Chapman in the 1820s could be in near perfect agreement with each other. The reason for this strange paradox was simply that Godman's viewpoint was, at bottom, not chemical but mechanical. His aim was to chart the workings of the always unknown "essence of life" in the transfer of impressions through the nervous system, and the viewpoint of the day was that such a tracing of activity would not suffice to remove the fundamental mystery from life.[49] But even though they insisted upon preserving

mystery, the proponents of sympathy generally held that their doctrine provided a "satisfactory explanation" of the phenomena; at
least one of them made it explicit that the doctrine was deliberately
offered as an antidote to the revival of "humoralism." [50]

For John Eberle, the doctrine of sympathy failed to make any
sense as an explanation. In an answer to Charles Caldwell, Eberle
countered that the idea that a phenomenon was "satisfactorily explained" by an abstract term which had no objective meaning was
"bordering very closely on the ridiculous." [51] Eberle, one of the
leading "humoralists" against whom Caldwell had spoken, could not
be satisfied with an explanation which did not include the mutual
influence of the solid and fluid parts of the body—that is to say,
chemical activity.[52] But while mechanical activity had been assimilated to the vitalistic viewpoint, it was still thought, as the paleontologist Richard Harlan argued, that the doctrines of chemical life
"have a natural tendency to debase man from his dignified and
compound condition of soul and body, to a mere complex chemical
machine." [53]

Furthermore, not only were such explanations undignified, but
according to an increasingly influential school of thought they were
entirely unnecessary. An unknown contributor to Elisha Bartlett's
Monthly Journal of Medical Literature related with a condescending
touch of amusement some of the errors of the prior generation. His
misguided predecessors, he said, had done a great deal of useless
work in the belief that much benefit could be derived from a
knowledge of the modus operandi of medicines. But although
curiosity might be gratified by a solution of this difficulty, the writer
did not see how it could aid medical practice. All that was required
for successful therapeutics was that the effects of the medicines on
specific diseases be known, and this knowledge could be gained only
by observation. Again, he contended, the medical world had often
involved itself in disputes over what constituted the essence of life,
and the profession had been much amused by the wildest speculations:

At one time a *vibration* of atoms agitated the public mind, and then it
was *alarmed* by the *apparition* of an aethereal spirit. Now it was a cause
which exercised power over matter, and anon, was transformed into an
effect, resulting from peculiar combinations of this same matter. These
dreams were severally insisted upon with zeal and obstinacy, as if es-

tablishing a truth which would have a powerful bearing on medical practice.

The writer pointed out that misconceived and ill-founded speculations had been a peculiar mark of the previous century, and they had continued until very close to the time in which he was writing. Such nonsense, he was happy to say, was no longer tolerated by the medical profession—or indeed, by philosophers in general. ". . . the more correct views of the present day," he rejoiced,

of the dependence of facts, and of the nature of cause and effect, readily concluded that this was an assumption altogether gratuitous—that life was one of nature's mysteries, and that, if solvable, its discovery would shed but little, if any light upon the nature of diseases, or enable us to locate them with more certainty. Men have learnt that such exercises of the imagination, with the barbarous experiments which have been resorted to for their illustration, however they may gratify the vanity of their promulgators or prove a monstrous development of the organ of Destructiveness, add but little to the mass of real knowledge, and indeed often serve to retard and obstruct its march. Diagnosis is the great secret of perfection in our art, and by the aid of those fixed and unchangeable laws, by which we know the animal economy to be governed, we are assured of the existence of a vital principle on which these laws depend.

Furthermore, he argued, it was not at all necessary that one be able to view the laws abstractly apart from the phenomena of the organic system in order to reap all the benefits of the knowledge that one did have. The vital principle assured physicians that there was a sensibility which pervaded the system, and which pointed out the particular organ affected and indicated the kind and degree of action occurring.[54]

The claim of the anonymous physician was simply that the inability to understand causes need not interfere with practical matters like diagnosis and treatment of disease. And as long as diseases were equated with symptoms, this kind of argument seemed reasonable enough, and sufficiently modest.[55] One could, however, make a great deal more than this out of the analogy between vital forces and ultimate physical principles. As early as 1826 the Philadelphia physician Samuel Jackson indicated in what direction a consistent pursuit of the analogy might lead. He began by agreeing quite readily with Chapman that the knowledge and means of research of the time were not developed to the point where scientists could ascertain with certainty the remote cause of life. Happily, it was not necessary that such a cause should be known; it was only necessary for men to know that organic actions were directed by established laws that always operated the same under similar circumstances. The *only* business of the medical philosopher was to determine what those laws were. And fortunately for him, those laws were within the domains of knowledge, cognizable by the senses, and could be made the subject of experiment. The examination and elucidation of such laws did not require a knowledge of the remote cause of life and probably, he thought, would receive little aid from its discovery. "A subject so abstruse and incomprehensible," Jackson concluded, "may then be dismissed as vain and frivolous."

Jackson's appeal to ignorance differed only slightly from Chapman's; the two agreed that there was an unknown vital power beyond the phenomena, and both thought about it in terms of an analogy with the ultimate physical principles. But Jackson's thinking went a step beyond that of Chapman:

A principle or power may exist, as the first cause of organization; but this is different from the vital properties, which are attached to the tissue, and belong to them, only as a result of their organization. . . . Viewed in this manner, there is probably but a single vital property which is irritibility.[56]

This statement by Jackson suggests that vitalism had had a treacherous ally in positivism. Ampere had used the positivistic approach to explain all the electrical phenomena then known, and in so doing he had made quite irrelevant the question of what an electrical current was. This is precisely what Jackson was threatening to do to the "vital principle." Very well, he seemed to be saying to the vitalists, so we admit there is an unknown vital principle and that it cannot be studied by our methods. But it really makes no difference; just as with the electrical current, the only thing that matters is the *property* of that principle, which can be studied by our methods. The vitalistic argument had not been refuted—indeed, it could not be refuted—but it had been made irrelevant, which was far more damaging to its standing.[57]

In making a distinction between principles and properties, Jackson spoke for the next several generations of physiologists, and he did the same in proclaiming that irritability was the relevant property. Irritability he defined simply as a property inherent in all organized matter by which that matter possessed the faculty of entering into action from the impression of stimuli. This was essentially the way Claude Bernard defined irritability in 1889, and with a few refinements it is still the way it is defined.[58]

The distinction that Jackson made was sometimes considered spurious by his contemporaries, for they saw very well that its tendency was to do away with vitality as a relevant constituent of physiological thought. An anonymous contributor to Silliman's *Journal* understood the danger inherent in assigning such a property to the elementary parts of animal and vegetable bodies. "These doctrines," he said, "adopted in their full extent, restore the dogmas of the metempsychosis, and the chances of Democritus, and, by vulgar induction, end in atheism." [59] Silliman, in a footnote to the article, suggested a better way to explain the "apparent animation" of inert matter. The microscope, he observed, had discovered innumerable animalculae in and upon almost everything. Therefore, it occurred to him that the observed animation might be due to the adherence of these tiny organisms, "whose origin is doubtless by the regular although singular process of life." [60]

Despite such objections as these, the distinction between principles and properties was obviously too fruitful to be ignored. Beginning with such a dichotomy, physiologists could consign to another realm the question of vitality and give their chemical explanations of the phenomena they were investigating. One could believe in a vital principle or not believe, as he chose, and it would make absolutely no difference in the way he did his work, or in the physical conclusions at which he arrived. As Richard Harlan, himself a staunch believer in final causes and in vitalism, noted: "To the closet of the metaphysician we would willingly consign the fruitless inquiry into the nature of final causes, and confine our researches to the study of accessible laws." [61] The next step, which no one in this period was willing to take, would be to declare that the study of the "accessible laws" could provide a complete understanding of nature. Someone did, however, draw one of the more obvious implications from the positivistic arguments that vitalists had been using. Elisha Bartlett, a French-trained physician and medical theorist, came nearer than any other American of his time to carrying these arguments to their logical conclusions. Science, he said, consists of ascertained facts or events, the whole classified and arranged. In case his meaning was not clear, he emphasized it:

These phenomena, thus classified, constitute, *not the foundation, and the materials, merely,* on which, and out of which, by some recondite process of the intellectual powers, called inductive reasoning, the science is to be contructed; they *are,* the science, *in themselves, wholly and absolutely.*[62]

Since all knowledge was incomplete and men had a natural desire to make it complete, they had tried to reach beyond the phenomena by using various "hypotheses." But all science, he said, "is absolutely independent of hypothesis." If one really pursued this idea, Bartlett realized that it had obvious implications for both final causes and the vital principle. The doctrine of final causes was certainly true—if one only meant it in the trivial sense that throughout the universe means were invariably and perfectly adapted to ends. But the argument of the natural theologians, which was an hypothesis drawn from the fact of adaptation, remained an empty speculation, outside the realm of certain knowledge.[63] In a later passage, applying the criteria of whether they consisted of "phenomena and their relationships, of an appreciable and positive character, ascertained by absolute and extensive observation," to the arguments in favor of a vital force, he concluded that there was no evidence of even the existence of such a thing. The whole doctrine, he said, was nothing but "physiological transcendentalism." And then he made the generalization toward which the whole positivistic argument had been leading: *"Life is the sum of the organization and its actions. This is all we know—this is all we can know, about it."* [64]

As the vitalists were anxious to preserve the body of man from the prying eyes and impious inquiries of speculative science, they were equally alarmed lest the earthly abode of man become subject to such scrutiny. One who insists upon wise contrivance of a benevolent Creator to explain the mechanism of a living organism generally finds, for the same reasons, that there is wise contrivance in the environment upon which that organism—be it man, reptile, or zoophyte—must depend. Hence, it comes as no surprise to find that in the second quarter of the nineteenth century precisely the same kind of argument was taking place in geology as in physiology, and with the same results. The earth was made for man, there had been purpose in its creation, and every topographic feature testified to the omnipotence of the Creator. While religion could with only minor adjustments accommodate the belief that the earth was formed in time, it could not with the same facility give up the proposition that the formation of the successive stages required the intervention of the Creator. Was there not, throughout all of geological history, an admirable relationship between the conditions of the earth and the dominant life forms found on the earth? A chemical explanation of geological origins which omitted teleology was therefore just as suspect from the viewpoint of finalism as was a chemical explanation of physiological functions which threatened the vital force.[65] And the most offensive part of geology, the historical aspect, was subject to the same positivistic criticism as physiological speculations had been. "It is not the science to which we object," wrote Jared Sparks,

but the theory; not the facts, but the speculations; not the realities, but the dreams. Geology, in its legitimate objects, is a science of observation and analysis, and should be confined within its proper limits. The structure, the gradual revolutions, and the component parts of the earth, are subjects capable of investigation, and have their utility; but inquiries about the original elements of chaos, and the formation of the globe out of those elements, are as preposterous as they are fruitless.[66]

Sparks was making the same positivistic appeal against speculation about geological origins as Chapman made against chemical speculations about vital forces. He would limit geology to the description of ongoing changes in the structure of the earth in exactly the same way Chapman limited physiology to describing and naming bodily processes. Chapman used the positivistic arguments to maintain the importance of an occult vital principle in animal life, and Sparks used them to keep his theological interpretation of creation safely out of reach of chemical or physical theory. Once again, as with Sparks' complaint and in most such cases, the offending science was chemistry. Ira Hill's theory, which Sparks had been protesting against, outlined a *chemical* process whereby the earth was formed "out of the original elements of chaos."

In the controversy over geological speculation, one finds the same elevation of a temporary condition of ignorance into a supposedly permanent one, as had occurred in physiology. J. G. Cogswell observed in 1822 that human curiosity is easily baffled and human science confounded by nature. "After all the researches, which have been made," he exclaimed rapturously, "what secret of hers has been discovered!" Though geologists had visited all the volcanoes of the world, for instance, and had meticulously examined the streams of lava flowing from them, they could still not say how the process of volcanic activity was carried on "in the great laboratories of nature." From the current inability Cogswell concluded that these matters involved "questions beyond the reach of human powers." He would limit the geologist to investigating "changes which have taken and are continually taking place on the surface of the earth by the agency of the different elements." [67]

A few years earlier William Maclure had outlined the legitimate objects of geology in much the same way in an address before the American Philosophical Society. "In all speculations on the origins, or agents that have produced the changes on this globe," he maintained,

it is probable that we ought to keep within the boundaries of the probable effects resulting from the regular operations of the great laws of nature which our experience and observation has brought within the sphere of our knowledge. When we over-leap those limits, and suppose a total change in nature's laws, we embark on the sea of uncertainty, when one conjecture is perhaps as probable as another.[68]

Maclure, like Sparks, was also a static creationist, and accordingly confined geologists to investigating "regular operations of the great laws of nature"—in effect, forbidding a scientific explanation of the origin of the earth, for as a recent student has observed of Maclure, natural laws and processes seemed to him essentially conservative in character.[69] Nature tended to maintain things in the state in which they were found, barring some kind of extra-natural creative act. One could not, for example, imagine a change from a fluid to a solid state without a corresponding change in the laws of nature:

The supposition that the earth was in a fluid state when it took its present form, leads to the supposition that it was always so; and that fluidity was the original state of the earth, kept so by all the general laws of nature, all of which general order and laws of nature must be totally changed before the earth would take a solid form.[70]

With his belief in the conservative character of natural laws, Maclure could use, as Chapman had, the appeal to ignorance in order to maintain his ideological commitments. Later geologists would, of course, proceed along exactly those lines of investigation that Maclure had said was open to them. The only difference between uniformitarianism and the positions of Maclure, Sparks, and a host of others in the nineteenth century was that the uniformi-

tarians declared these processes a sufficient explanation in themselves—as physiologists had declared the study of the properties of
vitality to be sufficient. Edward Hitchcock, in 1829, expressed his
reason for endorsing the Cordier theory of the earth with an
appropriate quotation from Scrope's work on volcanoes:

They have one immense advantage over most, perhaps over all, of the
hypotheses that have been as yet brought forward to explain the same
appearances, and which speaks volumes in their favor; and this is, that
they are still in operation; with diminished energy it is true, but this is
the necessary result of their actions.[71]

Thus, in both geology and physiology the positivistic arguments
of finalism were turned against the finalists. In both cases those
wishing to provide a completely natural explanation had merely
taken their opponents at their word. In physiology it was accomplished through a distinction between vitality and vital properties,
and there was no longer any need even to consider a "vital force" as
a relevant variable. And in geology a distinction was made between
origin and process; geologists discovered that the laws of the processes could explain as much as was relevant about origins. The
finalists had thought to hold their ideological commitments forever
inviolate from the encroachments of chemistry by the insistence that
science was bound to "positive knowledge." But in both cases the
new views were developed out of, and in response to, this positivistic insistence, which finalists wished to use for their own ends. As
the next generation of scientists told the finalists, the positive knowledge their science was allowed was really quite enough.

∾

The Inductive Process and the

Doctrine of Analogy

Scientists could not for long ignore the fact that they were bound, in some measure, to take account of that which they could not directly observe. The newer scientific interests of the period—historical studies like geology, or the study of the imponderables, such as light, heat, and electricity—obviously demanded some kind of inference that went significantly "beyond the facts." Conventional formulations of the inductive process were of little help for these interests, for they were formulated with reference to an earlier stage of science. Generally, the inductive models were quite nicely calculated to derive answers to problems of the science of Bacon's day, but certainly no later than that of Newton. Scientists of the early nineteenth century, in a word, were erecting a philosophy of natural history for a world in which analysis was quickly replacing classification, in which construction was replacing discovery.

According to the usual interpretation of inductive certainty, it was to be extended only to the act of classifying those materials immediately at hand. The Scottish philosophy, it will be recalled, offered a convenient justification for belief in the validity of one's classification *as a simple description,* for "man is so constituted" that he could not doubt the reality of that which presented itself to his senses. Therefore, taxonomic judgments taking the form of "this C resembles A more nearly than it does B" could not be doubted, for—so the opinion went—this was merely a "complex" sense impression. But once the judgment was extended to "all C's resemble A more nearly than they do B," the received Scottish philosophy provided no really satisfactory justification. Such extensions of the inductive generalization were—with marked reluctance—admitted into science as "theories"; but in careful speech they could not be given the more honorific title, "laws." The greatest difficulties in the inductive problem, however, could be overlooked as long as the generalization could be regarded as a simple probability consideration concerning the next items in the series to be examined. One could always regard such conclusions as tentative, and he could hopefully await confirmation or rejection, which was sure to come soon by the ordinary processes of science. But if the next items in the series were not to be directly examined—if they were beyond human powers of observation, or if they were either far in the future or far in the geological past—the problem became entirely different. Could one assume that they, too, would conform to expectations, even when he had no way of directly testing the assumptions? One common reaction to such a perplexing situation was discussed in the preceding chapter. One simply drew a line at those points beyond which direct observation could not be expected to go, and declared that this marked the limits of science. Others, not quite so willing to adopt this defeatist attitude, began searching for ways to broaden the range of their philosophy.

Newton, of course, to whom American scientists frequently appealed, had offered as one of his rules of philosophizing a rule of thumb concerning inductive generalizations: "The qualities of bodies which admit neither intension nor remission of degree, and which are found to belong to all bodies within the reach of our experiments, are to be esteemed the universal qualities of all bodies whatsoever." (Rule III) But Newton himself had realized that his experimental method could not assure truth—it could not assure that the very next experiment would not invalidate an apparently well-founded generalization. And to minds in need of certainty, a type which seems to have been well-represented in nineteenth-century America, such a provisional basis for science would never be satisfactory. Consequently, when American scientists did appeal to this particular rule, they were much more likely to fasten upon another part of it, where Newton admonished them not to relinquish the evidence of experience for the sake of "dreams and fictions of our own devising." With no justification provided for it, Newton's own rule seemed to them suspiciously like one of those "dreams and fictions."

Neither American philosophers nor scientists were much concerned with fundamental epistemological problems of the type David Hume had grappled with. The Scottish philosophers had met the enemy there and had vanquished him to the satisfaction of the Baconian mentality. That part of Hume's argument which was not simply dismissed as the deranged wanderings of an impious sophist was easily restated for accommodation within the Scottish philosophy. The Humian argument on the cause-effect relationship could, for example, be put to valuable use. Scottish philosopher Thomas Brown had framed the syllogism that most American thinkers accepted: As Hume has shown, there is no power within the natural process to connect cause and effect; but immediate experience indicates that there is a constant connection between cause and effect; therefore, there must be a Power outside the process to account for the connection. With the aid of Scottish realism, Hume's demonstration could be thus neatly incorporated into the natural-theological framework of the nineteenth century.[1]

But the ready dismissal of such fundamental questions could certainly not be confused with an answer to the question raised by the unobservables. Whether one desired "ultimate" explanations or whether one merely wanted to keep within the observable sequence of events as the more positivistic were arguing that science should do, the problem was fundamentally the same—how to be secure in their own minds without the need for endless experiments or observations. In 1831 Francis Wayland, a leading American representative of the latest generation of Scottish philosophy, offered the scientists a justification for Newton's rule as a solution to their problem. Wayland, at the time of his Phi Beta Kappa oration "The Philosophy of Analogy," was the president of Brown University. He had previously studied medicine, had been a minister of the gospel for several years, and had briefly held a professorship of mathematics and natural philosophy at Union College. He had, one could say, investigated nature as both scientist and theologian, and in 1831 he was speaking in both capacities. Starting with a *tabula rasa* theory of the mind, Wayland postulated in man a "universal appetite for knowledge, which, by a law of his nature, grows by what it feeds on." The universe, by Beneficent Providence, was designed to serve both as fodder and as appetizer for the human mind, for it was made to correspond to the mental appetite in all respects. This much was a most conventional analysis drawn from the Scottish philosophy. But Wayland realized that neither the appetite nor the relation of correspondence could be the source of knowledge about a part of the world not immediately available for inspection—and he was aware that it was this part of the world that was of most consequence for a philosophy of science. Some tie had to be provided between the mind and the world outside the range of perception. Wayland's solution, which depended on the faculty psychology of Scottish philosophy, was that man was to make the leap from the known to the unknown by the God-implanted faculties of the mind, "by the exercise of which he is able to discover that truth by which his desires are gratified and his intellectual happiness created." [2]

Wayland went a great deal further in explicitly rejecting contemporary Baconianism than most of his colleagues would have been willing to go. Laws of nature do exist, he insisted, but contrary to the prevailing ideas they are not visible simply by inspection. The changes going on in the universe—the changes that scientists sought to understand—were seen only as results; neither the laws nor the processes of change were subject to immediate observation. In trying to understand his world, man had two methods of thought open to him—demonstration and induction. The former, he said,

proceeds from self-evident principles to the most complicated relations. Its sphere is the science of quantity, and within that sphere, its dominion is absolute. . . . The other instrument is induction. By means of this we commence with individual instances, and, by comparison and classification, arrive at laws more and more general. . . . The difference between these two processes is this. The one proceeds from self-evident truth to its necessary results; the other from known effects to their actual antecedents. [3]

It was with the second of these methods that the scientist was concerned, Wayland noted, but there were special difficulties in his use of induction for interrogating nature. When the scientist asked a question of nature, she always replied to that single interrogation alone, and the answer never went beyond a simple "yes" or "no" to the question of existence. Furthermore, since the negatives in nature's language were as difficult to interpret as the affirmatives, one might spend years in experimentation only to receive a "no" in answer to his question.

Wayland, at this point, had seen the inadequacy of the usual formulation of Baconian methodology. In effect, he had said that there was a great deal more to science than observation and induction as it was ordinarily understood. In reflecting on the problem of knowledge, it seemed to him that if one were possessed of no other means of discovery than the strict exercise of the reasoning faculties, any additions to the body of knowledge would be merely accidental. "Demonstration and induction never discover a law of nature," he explained, "they only show whether a law has or has not been discovered." To aid the scientist in discovering a law of nature, two added elements were needed—skill in interpreting nature's answer, and skill in asking the question. This last was the science of analogy.[4]

Wayland nowhere defined what he meant by analogy: it is likely that he knew his hearers were sufficiently familiar with it not to need a definition. He preferred to illustrate his meaning, appropriately enough, by an analogy:

Suppose I should present before you one of the paintings of Raphael, and covering by far the greater part of it with a screen, ask you to proceed with the work and designate where the next lines should be drawn. It is evident that no one but a painter need even make the attempt; and of painters, he would be the most likely to succeed, who had become best acquainted with the genius of Raphael and had most thoroughly meditated upon the manner in which that genius had displayed itself in the work before him. So, of the system of the universe we see but a part. All the rest is hidden from our view. He will, however, most readily discover *where the next lines are drawn,* who is most thoroughly acquainted with the character of the Author, and has observed, with the greatest accuracy, the manner in which that character is displayed, in that portion of the system which he has condescended to reveal to us.[5]

Thus, an analogy, according to Wayland, had to be based upon a belief that the unknown would correspond with the known. In other words, if A and B were alike in characteristics P, Q, and R, we could expect them also to be alike with respect to characteristic S. This was certainly not profound doctrine, and Wayland had as yet done no more than state it as an assumption derived from the prior assumption of the uniformity of nature. But Wayland's statement of this principle was important—because he realized that it carried him beyond the bounds of the Scottish philosophy. There was no faculty of the mind to which a belief in analogical evidence could be referred, for an analogy as Wayland understood it meant more than a sense impression or a comparison which the Scottish philosophers characteristically confused with a sense impression. The heart of Wayland's doctrine was the predictive element—the transition from the observed to the unobserved—and this quite obviously had to do with nature rather than with minds. Seeing that in this case some assumptions had to be made about the ruling principle of nature itself, Wayland made two "self-evident" principles explicit:

First. A part of any system which is the work of an intelligent agent, is similar, so far as the principles which it involves are concerned, to the whole of that system.

And, second. The work of an intelligent and moral agent must bear, in all its lineaments, the traces of the character of the Author.[6]

These two principles were the vehicles by which the very different problems of dealing with the *unobserved* and dealing with *unobservables* were made to appear the same. The first principle states that one can reason from what has been observed to what has *not yet* been observed; the second states that one can reason from what has been observed to what *cannot* be observed. The justification in both cases is the same: everything in nature is the work of the same "intelligent agent" and therefore "must bear" the lineaments of that agent. The consequent identification of the two problems was characteristic of the thought of the early nineteenth century, and it was responsible for a great deal of the confusion in the use of the word "analogy."

Wayland, in postulating his two principles, did not answer the Humian question in Hume's own terms, but rather dismissed it as irrelevant. In this respect he was quite characteristic of the regnant Scottish philosophy. In speaking of "self-evident principles" and "known effects," he had already taken for granted the fundamental question at issue, and in reasoning from cause to effect (starting with the self-evident principles) he had broken one of the most important rules of nineteenth-century Baconianism. At bottom, his position was that men could know and explain nature because a benevolent Deity assured uniformity in all its parts, and since this was so, there was no reason for deeper inquiry. Science, then, was to rest upon the axiom of God's goodness, and, reasoning from this axiom, it seemed obvious to Wayland that an analogy—that is, an observed similarity in the effect—would assure that there was also an analogy in the cause.

Wayland's axiom concerning the benevolent God was not the same as the assumption that "nature is uniform" which later logicians announced as the foundation of the inductive process, although it did serve the same purpose. However broad and general it might be, "Nature is uniform" is a statement of fact. Since this is so, it is itself an inductive generalization, the truth of which can only be established by an appeal to the facts.[7] But Wayland's axiom, insofar as it carried any weight at all, was basically neither descriptive nor predictive, but normative. The work of this intelligent and moral being "must bear" traces of the character of the Author; and for Wayland this meant that there *must be* a unity of plan. The difference between these two formulations is mostly psychological —those who followed Wayland's axiom would have an ideological stake in finding uniformities, and they could not be expected to insist upon a very high degree of uniformity before they began projecting it into unknown areas. In other words, it was admirably designed to promote loose reasoning and hasty generalization. Where previously the process of confirmation had seemed intolerably endless, Wayland would dismiss it as largely unnecessary.

The difficulty in Wayland's formulation was simply that the question of the degree of resemblance required was left completely open. According to the theory at least, there could have been no way to evaluate the difference between (1) two phenomena bearing only a remote and chance resemblance to each other, and (2) two phenomena differing only in time. For example, that dogs and cats were revealed upon inspection to have four legs each, fur, claws, teeth, and other shared characteristics, would have been as valid a resemblance for reasoning as was the fact that two chemical operations differing only in the time of their performance produced equivalent results. While it would certainly be absurd to suppose that Wayland did not perceive a difference in value, the point I wish to emphasize is that such a recognition could not have been derived from his theory, but would necessarily have come intuitively, in spite of the theory. This was not a personal failure of Wayland's; quite to the contrary, his manner of solving the Humian problem made the omission inevitable if the integrity of the fundamental assumption was to be maintained. Once God's benevolent interest in assuring uniformity was questioned, even concerning the smallest detail, the door to intellectual insecurity would be opened wide

again. Furthermore, Wayland was far from being alone in his position, for it was a serious weakness shared by American scientists
both before and after the classic statement. The assumption behind
all Wayland's thought was that since a benevolent God assures
uniformity in all the various parts of His universe, that universe
would conform to our expectations in every particular. This postulate was raised to the status of an analytical truth by Wayland, and
from this it followed that analogy was not only a legitimate mode of
reasoning, but was to be a certain basis for the entire inductive
process. Newton had said, in his authoritative second rule for
philosophizing, that "to the same natural effects we must, as far as
possible, assign the same causes." [8] Wayland, in elaborating his
philosophy of analogy, substituted "like" for "same" and neglected
Newton's limiting phrase "as far as possible." These two small
differences were of monumental significance for nineteenth-century
science in America.

Wayland was by no means the first to introduce analogy to
science. His treatise receives attention here simply because it was
the first systematic exposition of analogical thinking during the first
half of the nineteenth century. In terms of the claims made for
analogy, he was a great deal more modest than his predecessors; nor
is his as extreme as claims made by later theorists.[9] In one place,
Wayland distinguished between the methods of discovery and of
proof, and, at first glance, he seemed to reserve the latter for
induction and demonstration. Yet even for Wayland, analogy was
more than a convenient tool to guide one's thinking, since later he
suggested that lack of analogy might be used to *disprove* hypotheses. Here again, he was quite close to the practice of modern
science, for there is no doubt that under ordinary circumstances no
hypothesis will be admitted that does not fit previous experience.
The difference is that with modern scientists the restriction is only a
methodological rule of thumb which has been broken in a great
number of significant cases, while with Wayland it was deduced
from what he thought was the necessary structure of the universe.[10]

Virtually any American scientist of the early nineteenth century
could serve as an illustration of the use of analogy in the Wayland
sense—that is, any except the large number who confined themselves to natural history description. But there were some who used
the method more overtly, for whom no long logical chain is

required to show that their reasoning "ultimately" rested upon analogy. Since these are the men with whom we are concerned, there is no better initial example than Benjamin Rush, America's best-known physician at the beginning of the nineteenth century, practitioner of "heroic medicine," and proponent of an influential single-entity theory of disease. It is his comprehensive theory of disease that is of importance for understanding the uses of analogical reasoning.

Rush began the reasoning that led to his single entity theory with observations upon what were then considered different types of fevers. All the known fevers had in common the fact that they were invariably preceded by general debility; this he inferred from the various causes commonly assigned to diseases. Considering all of these causes, Rush decided that they had one feature in common: they acted by reducing excitement of the system, by the abstraction of stimuli, or by the excessive or unusual application of stimuli. All of these alleged causes could be reduced to the term "general debility." The existence of these varied sources of "general debility" allowed Rush to consider as the same such opposites as excessive cold or heat, famine or intemperate eating, the "debilitating passions" or the "stimulating passions." Since considered in this manner there was but one "exciting cause" of fever (some kind of stimulus, or the withdrawal of an accustomed stimulus), Rush thought it followed that there was but one fever, varying only in terms of different "states." He then drew nineteen points of analogy between symptoms of fever and convulsions in the nervous system. Because one marked symptom of fever, a flushed skin, was at that time associated with distended capillaries, Rush concluded that the convulsions of fever took place in the blood vessels; therefore, bleeding the patient until action was reduced to normal was the appropriate treatment for all fevers. But there was more yet to this singular chain of reasoning, for Rush moved immediately from a general theory of fevers to a general theory of disease. All diseases were preceded by debility and their essence was irregular action, or the absence of the natural order of motion produced or invited by predisposing debility. Therefore—unity of cause indicating unity of effect—there existed but one disease, consisting simply of morbid action or excitement in some part of the body.[11] According to this analysis, the physician's job was simple; he had merely to choose one of the several available methods of exhausting his patient and apply the method "heroically." For assistance he would need a good supply of leeches and "blister beetles," the only useful articles in the *materia medica* being emetics and cathartics.

This is probably one of the most striking examples in medical history of a consistent and inexorable progression through a series of analogies to a general theory. What makes it of even more interest is that Rush was taken seriously by a large number of physicians in England and America during the opening years of the nineteenth century. The diagnostic simplicity which would allow the physician to cut through the frightening array of nosologies and the therapeutic simplicity which seemed to justify casting aside the cumbersome *materia medica*, which was largely worthless at any rate, must have been appealing to the medical world of that time.

It will not do simply to dismiss Rush as a weak thinker or a hasty generalizer. Obviously a man who could pursue the intricate series of analogies shown above—analogies based upon careful observation and the most up-to-date medical knowledge of the time—was hardly intellectually under-privileged. Nor was he overly quick to generalize, for if to Wayland's doctrine of analogy one adds a further assumption, all the results follow logically. The required assumption is the familiar one that constant succession in time is identical to the cause-effect relationship. Rush had observed that disease is invariably preceded by general debility; the general debility could then be said to be the proximate cause of the disease; if unity of cause indicated unity of effect then obviously there was but one disease manifesting itself in different "states." The added assumption required in this chain of reasoning was an easy one for Rush to make, for with but a few isolated exceptions the scientists of the early nineteenth century were convinced Humians in this respect if in no other. Attesting to the pervasiveness of this combination of empiricism and rationalism was the fact that while Rush's specific conclusions were attacked from time to time it was nearing mid-century before anyone took exception to the mode of reasoning he employed in reaching these conclusions.[12]

Such a convenient tool of reasoning as the analogical method could not be expected to remain confined to medical matters. A happy analogy occurred to the distinguished naturalist James D. Dana as he was reflecting upon animals in which the phenomenon of alternation of generations occurs—that is, in which a fully developed individual appears only once in every two generations. This, he thought, was not really so strange as had been supposed, if one only considered the normal course of plant reproduction: the seeds produce leaf-individuals, which yield bulbs or buds becoming flower individuals, and these produce seeds. This was precisely analogous to the process whereby the egg produced polyps, the polyps, bulbs that developed into Medusae, and Medusae, eggs.[13] This particular analogy, Dana concluded, was merely another instance of identity in the laws of growth in the animal and vegetable kingdoms, which, of course, signified identity of plan, and, by inference, the same Planner for both kingdoms. In the 1837 edition of his mineralogy textbook, Dana also followed the current custom of trying to force upon minerals the classification that was then being used in zoology and botany, complete with Latin terminology and all the trappings of binomial nomenclature. By 1854, however, realizing that this was perhaps pushing analogy a little too far, he adopted a different classification for his new text revision.

A year before Dana's paper on reproduction appeared, the mathematician Benjamin Peirce described to the American Association for the Advancement of Science an analogy in botany and astronomy developed by the Reverend Thomas Hill of Waltham, Massachusetts. Hill had noted that the angles which any two successive leaves, viewed from above, made with each other formed a numerical series in which the first two fractions had the numerators each one, and the denominators differing by one, and the terms of any other fraction were formed by adding those of the two preceding, as: 1/3, 1/4, 2/7, 3/11, 5/18, 8/29, and more. Turning to the heavens, Hill then divided the year of Uranus by that of Neptune, Saturn by Uranus, Jupiter by Saturn, and so on, and found a series: 1/2, 107/305, 21/52, 32/85, 53/131, 17/32, 8/13, and 22/57. Hill pointed out that this was very close to the series found on twigs and concluded that both the astronomical and the botanical problem had been solved by the same arithmetical law. "The botanical problem is," said Hill,

to distribute the leaves, buds, petals, etc., of plants in such wise as to secure a graceful variety of symmetry. The astronomical problem is, to proportion the years of the planets in such wise as to render the conjunction of any considerable number a rare occurrence; to secure, that is, the system from too great mutual interference, by keeping the planets scattered round the sun.[14]

Peirce agreed with Hill's judgment that one could draw from these facts evidence that one Mind created the universe. Science, once again, had demonstrated that God was the great mathematician.

The point that makes these otherwise trivial papers of such interest for an understanding of the science of their day is that neither Hill, Dana, nor any of the other scientists who searched nature for analogies were merely pointing out interesting parallels; nor were they primarily engaged in isolating variables for further study. Quite to the contrary, their assumption was clearly that in finding two unknown processes operating alike they had achieved a scientific explanation of those processes, and there was accordingly no need to go further. Dana's undoubtedly interesting analogy, for example, did not lead him into an investigation which might throw light on the difference between the Medusae and other forms of animal life; neither did it lead him into drawing further conclusions which could connect the Medusae with plants. And, by the same token, neither Hill nor Peirce attempted to go beyond the analogy Hill found between leaf placement and planet distribution. That certain fundamental differences between plants and planets might obviate any analogies in comparison did not occur to them. Why should they question further, since analogy *in itself* provided an explanation that accommodated an a priori conviction that an omnipotent Deity had been the Creator of both? Such differences as might be obvious in the makeup of various natural phenomena were thus irrelevant, it being the scientist's task to discover the symmetry, order, and harmony of God's universe. Pious scholars enthusiastically embraced the doctrine of analogy as a sophisticated scientific method which could best accomplish this purpose. By merely pointing out analogies inherent in natural phenomena they had already "explained" these phenomena in the only way they thought science could explain such things—by reference to the purpose of the Creator. Every instance of analogy discovered in nature doubtless strengthened their belief in a God-ordered world, and in turn reinforced their attachment to the analogic method. Even the "positivistic," who insisted that science could not "explain," could accept "analogical explanation" because it in fact offered, not an explanation within the process, but a teleological explanation.

Examples like these could be multiplied indefinitely. Seriously intended work was done in every field of science, based upon nothing but the loosest and most circumstantial analogies. The entire structure of phrenology, for example, was based upon an elaborate series of analogies, and even the opponents of phrenology rested their argument on the lack of analogy in certain areas.[15] Because the analogies demanded it, an American astronomer, Daniel Kirkwood, was led to postulate the existence of two new planets, one between Mercury and the sun, and one between Mars and Jupiter. Unembarrassed by the fact that his analogy demanded that the latter be so large as to be readily observable, he suggested that it was broken. And such was the persuasiveness of his logic that most of his American contemporaries regarded his work as having confirmed Laplace's nebular hypothesis. The British Association, however, although it treated his work with respect, did not think that the extreme claims would be accepted in England.

The extreme lengths to which a frantic search for analogy unguided by any further theoretical considerations could be pushed was exemplified by Amos Eaton in two articles in 1829 and 1830. In the first of these papers, Eaton marshalled a number of examples to demonstrate that the number five was the "most favorite number of nature." Half the known plants had their parts of fructification in fives, or a number which is a product of five; there were rays of fives or their products in coral and in animal radiata; men had five toes, five fingers, five senses, and so on. Furthermore, five principles "constituted" the highest vertebrate, man—inert matter, the attractive principle, vitality, the sentient principle, and intellect.[16] Fortified with his discovery that nature preferred to work by fives, Eaton next applied these considerations to geology. "I intend to demonstrate," he said, "that all geological strata are arranged in five analogous series; and that each series consists of three formations." Following this declaration he listed a series of unrelated formations, jumbled as to locality and sequence, that happened to agree more or less in some of their physical characters.[17] Eaton was not, however, allowed to build his structure unchallenged. G. W. Featherstonehaugh, in one of the most savagely critical reviews of the period, commented that if any of Eaton's pupils "should take it into their heads to add new stories to the grotesque edifice he has commenced, the whole world will by and by think us demented." Werner, the father of the Neptunist school of geology, might be singularly happy in two respects, the reviewer maintained—"one, that he did not outlive his reputation; the other, he died before Mr. Eaton applied his views to American earth." [18] Featherstonehaugh, however, did not object so much to Eaton's analogies as he did to his weird terminology, his classification based upon mineral content, and the extravagant claims he made for his own discoveries.

While Eaton and his fellow scientists were boldly enjoying the license offered them by the philosophy of analogy, the theorists had not neglected the subject. Samuel Tyler, outstanding exponent of Baconianism in the period, had argued that all the evidence on which the inductive process was conducted could be divided into analogy and identity, the former being defined as any real resemblance less than identity. He did not try to justify analogical thinking, beyond repeating the common Scottish philosophy formula: if we see an object, we cannot but believe in its existence, so if we perceive an analogy between phenomena, we cannot but believe that they are produced by a similar or common cause. Why this conviction is produced, Tyler thought, was a question "beyond the boundaries of philosophy." The existence of this conviction in the minds of all men obviated any further justification.[19]

The praise of analogical thinking had reached such heights that all that remained was for someone to take the final step—identify the use of analogy with the very method of science, induction. This conclusion was reached in 1848 by James D. Whelpley. Whelpley, it will be recalled, had argued that there were phenomena of interest to science to which the test of experience in the Baconian sense obviously could not be applied—we could never feel a particle of matter or see a pulse of light, for example. It was cases like these, Whelpley argued, that made one see that the inductive process was really the same as analogical reasoning—for the desire for certainty in man had to be fulfilled. It was only the method of analogy that could lead one to knowledge of the fundmental facts in the universe, by leading him through a consideration of that which was known to that which was unknown. In considering an atom, he said, one's object is to unite all the substantial forms and qualities suggested by sensible properties and appearances of larger masses which yield to the inspection of the senses. When one has made the analogical transfer from visible masses of the substance to the smallest possible mass of the same substance—a particle indivisible and indestructible—he has reached the idea of an atom of the substance. It was the same with a man. One had only to become familiar with all external properties, then conceive of him as an individual, a unity, the simple and undivided source and cause of all the appearances that invest and define him. Then if the spiritual idea of a man was in conformity with sensuous knowledge of him, it was true.[20]

Whelpley's juxtaposition of the atomic theory of matter with the spiritual nature of man was no coincidence. It shows very clearly what kind of difficulties scientists became involved in when they tried to substitute analogical reasoning for the Baconianism they had found lacking. They had so extended the "scientific method" that it was applicable to any and every subject, but with the unfortunate result that it could be used to "prove" anything. Thus taken to its logical extreme, the "new" method was even less satisfactory than the old.

Whelpley's article of 1848 had been published in a scientific journal, and it was tailored for scientists, who, as observed before, at that time usually tried to be hard-headed non-transcendental Baconians. How far they had in reality drifted from their original commitment was illustrated in an article written by Whelpley the following year for the *American Whig Review,* of which he was then editor. The subject of his article was a highly appreciative analysis of the "Life and Writings of Coleridge," a man considered by Tyler to be an arch-exponent of transcendentalism and German mysticism. While celebrating the excellence of Coleridge, Whelpley took advantage of the occasion to make the following comments about scientific method:

The judgement operates by three distinct modes or faculties—as first, by syllogism; of which the determination of a species under its genus, etc.; second, by arguing from cause-and-effect—as that the same cause shall always produce the same effect; and lastly by *analogies*—as when we say, that the same order or system of things, discovers the same principle controlling them—a species of reasoning which has a double certainty and value, from its embracing the principle both of the syllogism and that of cause. Yet the miserable logic of the last century, warns us in a very effectedly wise style against the danger of too free a use of the argument of analogy. When one sees the greatest absurdities stilted along upon syllogistic and cause-and-effect argument—one's fear of too free use of analogy is very much abated. Not staying to develope [sic] the entire system of the logic of analogy, we need only advert to the fact that every successful scientific or psychological speculation will be found to rest upon it, and if any peculiarity of method can be attributed to modern logic as distinguished from the syllogistic of the scholastics, and the cause-and-effect of the mechanical deists, it is the analogic of the moderns, pre-eminent, as including and subordinating the others.[21]

Whelpley's version of analogy was a great deal different from that of Francis Wayland, although the two were based upon the same assumptions. While it would be difficult to argue that Wayland actually had in mind any more than what is now meant by inductive generalization, Whelpley would include all of logic within it and have logic be dependent upon it. The kind of reasoning he advocated was from effect to effect, or from particular to particular, with both cause and general law simply being assumed.[22] His transcendental orientation became clear as he summed up in a single paragraph the conclusions of all analogical philosophy:

That spirit is before matter in the order of being; that phenomena in perception, and laws and principles in intellect are true analogues of certain realities in universal nature; that as there is a particular life of the individual, this is only a spark from the universal life of the world; and as there is a rational soul of the individual, this is only a spark from the Universal Person, the I Am; that in the world is both appearance and substance, but that substance can be perceived only by appearance, and known only *through* intellect.[23]

The philosophy of analogy, it seemed, could be sustained just as well by a God within the process as it could by a God outside the process. Perhaps this is one reason that Tyler found it so necessary to campaign against German philosophy, transcendentalism, and the like, for, as Whelpley's example indicated, his own brand of Scottish philosophy was dangerously like it in a number of particulars.[24]

While the type of thinking represented by Wayland, Tyler, and Whelpley seemed to arise within science itself, rather than being foisted upon it by religion, the scientists were never averse to seizing all the religious advantage it gave them. As Peirce and Dana had recognized, their detection of analogies not only gave proof of a God-ordered world, but also demonstrated their own piety. For their part, the theologians could certainly not be expected to miss the opportunity being offered them. Theology and science came together on the question of analogical thinking through the medium of the doctrine of final causes, a concept that Bacon had wished to cast out of science, but which had been reintroduced as a principle explanatory concept by the end of the first quarter of the nineteenth century. The philosophy of analogy could be gloriously extended and apparently justified through the postulate of final causes. The doctrine was of particular importance in the life sciences and could be seen in the physical sciences only in the dissatisfaction of certain thinkers with the explanatory basis of their discipline. Denison Olmsted was particularly disconcerted because his science (physics-chemistry) could not attain to the same satisfactory explanations as biology. When chemists tried to account for the known facts, they found it impossible to assign an adequate reason for any chemical action. All they could do in this situation, said Olmsted, was to "arrange such phenomena of this kind as are analogous into separate classes." [25] And, as observed earlier, Olmsted was evidently speaking for most of his contemporaries when he argued that explanation by reference to a higher law was a tautology. The doctrine of analogy was not quite so useful in chemistry, for there were no obvious purposes to which the analogies could be referred. Chemists, working with the "concealed and internal operations of the machine," had more difficulty in tracing a purposeful design than the natural philosophers or natural historians had with the mechanisms they described. This, perhaps, explains why there were so few efforts to derive a natural theology from chemistry, and why those few efforts were generally unconvincing.[26]

In other areas, as indicated in the preceding chapter, the doctrine of final causes did have great utility; in the hands of a naturalist like Cuvier it had produced some impressive results. Even such a skeptic as Cabanis had to confess his inability to do completely without such a principle, however abhorrent it was to him. "I consider, with the great Bacon, the philosophy of final causes as sterile," he said, "but I have elsewhere acknowledged that it was very difficult for the most cautious man never to have recourse to them in his explanations." [27] William Whewell, like Cabanis, recognized that Bacon had compared final causes in physics to vestal virgins devoted to God and therefore barren. Because of Bacon's unfortunate simile, Whewell acknowledged that the reference to final causes was sometimes spoken of as "unphilosophical." But with reference to that simile, Whewell said:

if he had had occasion to develope its bearings, full of latent meaning as his smiles so often are, he would probably have said that to those final causes barrenness was no reproach, seeing they ought to be not the Mothers but the Daughters of Our Natural Sciences; and that they were barren, not by imperfection of their nature, but in order that they might be kept pure and undefiled, and so fit ministers in the temple of God.[28]

Bacon was thus saved.

American scientists quite generally took the side of Whewell and Cuvier in the argument over the use of final causes. Richard Harlan, for example, reflected that it might be adopted as a principle

that the organization is universally in relation to the mode of life; and in consequence of their reciprocal influence, all the various parts of an animal are so closely connected with each other, that this relation can be traced even in the most minute particular.

Harlan forthrightly connected adaptation in the animal economy with natural theology. Comparative anatomy was of great interest to the natural theologian, he said,

who discovers in the modification of structure, according to situation and circumstances, and its constant relation to the wants, habits, and powers of animals, the strongest evidence of *final* purposes, and therefore the strongest proofs of an intelligent first cause.[29]

The belief in final causes meant to those scientists that there was an invariant relationship between structure and function—an analogy, that is to say. That a general relationship of this nature existed could hardly have been denied by the most determined skeptic, but by these scientists the doctrine was used to make analogy a substitute for experience. Until the 1840s, this kind of thinking dominated physiology, in which it was assumed that because of the necessarily wise contrivance fitted for an end function could be best discovered by inference from structures rather than by experiment on living subjects. There were, therefore, very few physiological studies of a modern type in America before the 1840s.

"Habit," as one naturalist said, "in numerous instances, is an infallible index of internal arrangement." There was therefore no need to go through the tedious business of morphological investigation, requiring as it did a knowledge and dexterity not possessed by many naturalists.[30]

Such implicit faith in the analogies of nature could not have been sustained without the doctrine of final causes, and there is good reason to believe that it was brought back into science by the scientists themselves because of its obvious utility at the time. Although earlier instances appeared sporadically, final causes no more than analogy seem to have been in such vogue during the first two decades of the century.[31] At least one presbyter of the Episcopal Church, as late as 1822, saw the inapplicability of final causes for science, and furthermore he recognized that the argument from design for the existence of God was essentially different from what the scientist was doing. "Many final causes," the Reverend Frederick Beasley noted,

may have contributed to the formation of the same object [i.e., it may have many distinct uses]. . . . Final causes, therefore, while they furnish unanswerable arguments in proof of the existence of a Supreme Contriver, have nothing to do, except as motives influencing the mind of the deity, in the product of effects, and of consequence, do not enter into the views or occupy the attention of the philosopher, in his investigations of nature, whose province it is to trace the series of causes and effects, or in other words, afford solutions of the various phenomena presented to his inspection.

It would be difficult to find a clearer differentiation between the concepts of purpose and cause, and Beasley did not bother to argue his viewpoint, merely presenting it as a common principle. Beasley also, like Harvard professor Levi Hedge (who had revised his text the previous year), gave a definitely subsidiary place to analogical thinking. He would only grant it power to answer the a priori argument, and he cited Bishop Butler's work as an ideal use of analogy.[32] Analogy and the use of final causes in science (as distinguished from the belief in them outside of science) seemed to grow together beginning in the late 1820s.

The implications for religion in the doctrine of analogy were too obvious to be missed. As James Brazer observed in an article in the *Christian Examiner* in 1835, natural theology could now be placed upon grounds as truly "scientific" as physics or chemistry. Natural theology had suffered in the past by being based upon "certain metaphysical propositions, which are assumed as axioms," particularly the principle of causation, of which Brazer (as a good Humian) was pleased to admit man knew nothing. Since one's knowledge was necessarily limited to the succession of phenomena, one could not demonstrate the existence of God by appealing to the necessity of a First Cause. Fortunately, Brazer maintained, the question of whether there was a God was a question of *existence,* which meant that it was not a matter of logic but of fact. Since this was true, His existence was to be demonstrated in the same manner as any other fact. The metaphysical mode of reasoning had no more place in natural theology than the syllogistic modes had with the interpretation of physical laws. Brazer then sketched the familiar outline of the inductive process. It consisted of two parts, investigation of analogous phenomena and inference from them of a general fact or principle, which, being thus ascertained, might be applied in all analogous cases. This mode of reasoning, he thought, was plainly as applicable to moral subjects as to material. For example, the design in the universe was precisely analogous to design by human agency in man-made contrivances. On this basis, Brazer thought, the argument from design for the existence of God was unassailable.[33] The "scientific method" could also provide indisputable proof that man was destined for a future life. Since all the instinctual needs of the lower animals were provided for in the circumstances in which they were placed, we must infer "in respect to man, as in reference to all other beings known to us, that all his powers, all his capacities, all the inherent principles of his being, are indicative of his true condition and destiny." If any of these found no adequate objects here, as he thought they obviously did not, then "we must admit that man is destined to a sphere hereafter, where the essential principles of his nature may be recognized and fully satisfied." [34]

Thus, for a few years in the early part of the nineteenth century, scientists, philosophers, and theologians found common ground in identity of method. Each thought that the questions of peculiar interest to them could be answered to their respective satisfaction with the same "scientific method." It indeed seemed that both metaphysics and theology could aspire once more to the designation "science." As a matter of fact, since the possibility of any knowledge of nature was made to rest frankly upon the postulate of God's benevolence, it would have been possible to regard both metaphysics and science as special problems in Theodicy, subsumed under the *scientia scientiarum,* theology. These are, in fact, the very words used by one writer to describe the relationship between natural theology and the "other" natural sciences. He concluded of natural theology "that it is the queen of all the sciences except the revealed; that it is, with this exception, the true *scientia scientiarum.*" The writer drew this conclusion because natural theology was the most comprehensive of all the sciences—that is, its subject matter and the subject matter of all sciences were coextensive.[35] Brazer was thus perfectly correct in asserting a unity of method, and one could not help but agree with another commentator, who noted that in recent years science had become more, rather than less, friendly to Theism.[36]

Today one becomes a little puzzled upon reading all this. While the reason for theological interest is obvious, it is difficult at first glance to see why the analogical form of thinking should appear so attractive to scientists and philosophers of that time. It is not merely that in the mid-twentieth century we view analogy as a very primitive form of induction, useful for suggesting lines of inquiry but untrustworthy as a mode of drawing conclusions. More important than any such presentist considerations, the men discussed here were working three hundred years after the formulation of a rigorous scientific method by Galileo and over a hundred years after the brilliant example of Newton; they themselves sang the praises of Bacon—to whom their reasoning would have been alien—in virtually every paper they wrote; and only fifty years before, Hume had been thought to have given the death blow to argument by analogy. He had, in fact, rejected the very analogy which Brazer now thought provided an unassailable basis for natural theology. The problem, then, is not why these scholars were less "enlightened" than we, but why there was an apparent return to an earlier mode of reasoning. Why this apparent disregard of the past two hundred years of scientific thought?

Quite early in the period, reinforcing the observations made by Hedge and Beasley, a thorough denunciation of analogical reasoning per se had appeared. Written in 1824 by John Eberle, editor of the *American Medical Review and Journal,* the article was published as a book review in the first number of that periodical. Eberle's subject was the publication by John Bell of an annotated American edition of Combe's *Essays on Phrenology.* Eberle was critical of phrenology in general, but his most intemperate criticisms were directed at Bell's attempted analogical proofs for the truth of phrenology. "Analogy," Eberle argued, "is a dangerous mode of reasoning. According to our apprehension, at least, the 'analogies of nature,' here adduced, serve quite as well to prove that the moon is made of green cheese, as, that the brain is divided into different organs." Unlike many commentators in subsequent years, Eberle did not confine his rejection to the specific "false analogy" to which he objected. It was the method itself that he attacked, and he directly connected analogical thinking with an absurd form of the doctrine of final causes.[37]

Eberle's effort was the last of its kind to appear in scientific literature until the early 1840s. There were many, however, who continued to object to analogies in specific areas. Francis C. Gray, for example, pointed out that both Wernerians and Huttonians extricated themselves from difficulties not by close deduction but by "loose and distant analogies." In the mid-twenties, John P. Harrison's critique of phrenology argued that its entire basis was nothing more than an analogy "with which, as by a lever, the disputant on either side of an argument helps himself out of his difficulties." A year later, William H. Shaw deplored the false analogies in medicine from physics and chemistry. In the end, however, Shaw endorsed Chapman's doctrine of sympathy, which was itself an outstanding example of a loose analogy.[38] All of these confined their disputations to the rejection of specific "false analogies."

The only scientific writer since Eberle to take the argument a step further was Ebeneezer Emmons, who in 1834 complained that plant physiologists seemed to have established too much by analogy. Following Bichat in the belief that analogies could be drawn between the plant and the animal kingdoms "without any violation of the strictest rules of logic," they had studied the functions of the different organs in animals and then transferred those functions to vegetables. The leaves were thought to perform the function of lungs and physiologists discovered a double function in vegetables, notwithstanding the fact that there was not the vestige of a heart. The only analogies which it would be safe to admit, Emmons concluded, were very general. For instance, it was obvious to him that different varieties of both animals and plants required different foods, could survive only in certain climates, and had different kinds of circulatory apparatus.[39] Thus Emmons seriously questioned the legitimacy of analogical reasoning, but even he stopped short of condemning its use on epistemological grounds. Although the categories of analogies he would admit were so very general that had they actually been restricted to this their use would have been effectively destroyed, he still offered no theoretical distinction between the admissible and inadmissible analogy. Scientists apparently were incapable of making any such distinction. Why this should have been so is a question that must be explored, but first it is necessary to ask why this type of reasoning should have been so widely popular.

One obviously reason was the difficulties that had appeared in applying the pure Baconian method. Analogy now seemed to be no more than a logical extension of Baconianism; in fact, to people like Samuel Tyler it seemed to be implicit in this philosophy. Bacon himself had suggested that analogies were useful tools and had promised a more extended treatment of the subject; this was one of the many tasks that he never got around to completing. And as Whewell had shown, Bacon's objection to the final causes at the basis of analogical reasoning could be quite neatly explained away.

The main problem facing scientists at the time, as Whelpley suggested, was to find a way to deal with unobservables, and for this purpose the method of analogy was quite appealing. It was no secret that Ohm had been guided every step of the way toward the formulation of his law of the resistance of a conductor by following a consistent analogy between the flow of heat and the flow of electricity. It was the same with the wave theory of light. Thomas Young based the entire theory upon an analogy of the vibrating waves of sound, and his favorite illustration was of the mutual interference of waves in a river. The importance of the contributions of Ohm and Young can hardly be overemphasized, and if one looks at other sciences, similar examples are readily apparent. For example, the first clear recognition of folding found in geological literature was an article in 1825 by J.H. Steele. "It is impossible to examine this locality," he said,

without being strongly impressed with the belief that the position which the strata here assume could not have been effected in any other way than by a power operating from beneath upwards and at the same time possessing a progressive force; something analogous to what takes place in the breaking up of the ice of large rivers. The continued swelling of the stream first overcomes the resistence of its frozen surface and having elevated it to a certain extent, it is forced into a vertical position, or thrown over upon the unbroken stratum behind by the progressive power of the current.[40]

Throughout this paragraph, Steele was describing what was actually seen every year when the ice broke up on the river. Yet his conclusions applied not only to the river but to the formation of geological strata as well, and it is clear from the article that Steele assumed a necessary connection between the two phenomena. It is unnecessary to belabor the point that there was no way Steele could have viewed folding firsthand, and this was true of almost all phenomena of historical geology; some substitute for direct observation was essential if the subject were to be studied at all.

One of the most obvious and immediate successes of the analogical mode of thinking was Cuvier's application of comparative anatomy to vertebrate paleontology. According to his principle of correlation, Cuvier held that organs in nature did not exist separately but only as parts of complete living beings, and in these living things certain relations obtained which were fundamental to their mode of life. Feathers, for instance, were always found in birds and never in other creatures. Cuvier thought this principle was traceable in the structure and working of every organ and indeed in every part of every organ. Thus, given a feather, it was possible to infer that its owner had a particular form of collarbone, skeleton, mouth, structure of lung, method of breathing, and so forth. Cuvier and his disciples developed the great knowledge of organic forms required for tracing relations, and applied the method to fossil reconstruction. His pupil, Sir Richard Owen, took the method to its greatest heights by reconstructing a giant bird of a very aberrant type from a little bit of leg bone. When his reconstruction was proved accurate by subsequent discoveries it became very difficult to argue with his method.[41]

This same comment applies to the other examples cited above. There is certainly a sense in which success carries its own credentials, and the results of Ohm's, Young's, and Cuvier's works were among the most strikingly successful of the early part of the century. These examples, coming at the very time when a great number of scholars were beginning to feel the limitations of pure Baconianism, practically guaranteed an attempt to revive the older mode of thinking. Although Hedge in 1821 and Beasley in 1822 could maintain that analogy was little used in science, Whelpley in 1849 could distinguish the logic of the moderns from that of the "Enlightenment deists" by the fact that analogy was at the very foundation of nineteenth-century science. Such had been the impact of the analogic revival.

Hedge had attempted a formal distinction between analogy and induction; his failure is a clue to the reason why nineteenth-century scientists found it necessary to admit all analogies as logical equals, to make no distinction in advance between "good" and "bad" analogies. "Resemblance or analogy," Hedge said, "is an extensive principle of association . . . ; one natural scene suggests another; and one event or anecdote frequently brings another to our remembrance, by the similarity we observe between them." He began by admitting the difficulties in differentiating between analogy and induction, for every inductive process did commence with analogy. Nevertheless, he believed that there were two circumstancs which set them apart. First, induction reasons from several individuals of a class to a whole, and its conclusions are always general. In analogy, however, one argues from an individual being to another of the same class, and from one species to another. Second, the evidence employed in analogy is wholly indirect and collateral—the co-existence of two qualities in one subject affording no direct evidence of their coexistence in any other. But in the inductive process one has direct evidence that the property applied to a whole class exists in many individuals of that class.[42]

Hedge, it would seem, did not accomplish what he set out to do. The only solution to the problem he could make was to take for granted the very point at issue. What, in logical terms, constituted a class? What degree of resemblance had to be demanded before one could consider phenomena the "same" for purposes of reasoning? He did no more, in effect, than to say that there were good analogies and bad; the good were called "inductions" and the bad "analogies." Surely, as the reviewer of the first edition of his book noted, this was no help. "We have a vague impression, and the metaphysicians tell us, that analogy and induction are not the same," Willard Phillips commented, "but when they attempt to point out the difference, they confound them." [43]

It seems that even before the method of analogy became equated with the logic of science neither scientists nor logicians were able to formalize an adequate basis for separating the objects of the world into classes from which valid conclusions could be drawn. A close look at one very common analogy should serve to show the precise nature of their inability. The example, taken from Thomas Reid, is an argument for the existence of life on other planets. "We observe," said Reid,

a great similitude between this earth, which we inhabit, and the other planets, Saturn, Jupiter, and so forth. They all revolve round the sun, as the earth does; though at different distances and in different periods. They borrow all their light from the sun, as the earth does. Several of them are known to revolve round their axes, like the earth, and by that means must have a like succession of day and night. Some of them have moons, that serve to give them light in the absence of the sun, as our moon does to us. They are all in their motions subject to the same law of gravitation as the earth is. From all this similitude it is not unreasonable to think, that those planets may like our earth, be the habitation of various orders of living creatures.[44]

Reid, as was customary before about the end of the 1820s, had only claimed that the analogy he adduced was a probability consideration—an answer to the a priori argument. In view of the resemblance, it was "not unreasonable to think" that these planets might be inhabited—that is, there was no valid a priori argument why this might not be so. Reid's impressive list of analogies did, of course, demonstrate that the planets had a great deal in common. But the striking fact was the total absence on the list of any characteristics that had to do with the question at issue: the ability of a planet to sustain life. Relevant characteristics would have included such things as temperature, atmospheric qualities, and density. Apparently Reid thought that any similarities would do, for he never even suggested that other points of analogy might be more convincing.[45] Although neither Reid nor anyone else ever thought of applying the argument to show the existence of life on the tiny asteroids, a comparable list of similarities could have been drawn up for them; an impressive list could even have been made for comets.

With this last example before us, keeping in mind what has been said before, the necessary conclusion is that there was nothing essentially wrong with the tool of reasoning the nineteenth-century scientists rediscovered. The fault was simply that those who were using it were trying to get too much out of too little. All too often they acted as though they had finally refuted the Humian argument by discovering an analytic principle of induction.

The times cried out to the men of the early nineteenth century that there was a God and that he was a benevolent God. Since this was so, it seemed that His works must necessarily attest to His character.[46] From this point, the argument developed inexorably toward the conclusion that since "one mind created the entire universe" there must be harmony in all its parts. Starting with the dogma of orderliness and adding to it the commonsensical approach of the Scottish philosophy, it was difficult for scientists to escape the conclusion that things which *look* alike must *be* alike. Furthermore, and most importantly, there was no such thing as a *chance* resemblance. As a corollary to this, there was no such thing as a resemblance that did not matter. The world was one grand plan of creation, each part manifesting purpose and bearing the lineaments of its Creator. Given the concept of purposiveness, which no scientist of the period really escaped, the desire to see any resemblance as indicative of unity of purpose was irresistible.

A deep commitment to ideas like the above is the only factor that can explain the excessively naive reliance on analogical inferences that so many scientists of that period displayed. Why else would a trained investigator like Nathan Smith conclude that *because* both plants and animals had a cuticle, a cutis, and a skeleton, there must therefore be something in the animal economy that could be compared to the radicles of vegetables? This necessary inference, as Smith thought it was, lay at the center of his argument that digestion was accomplished by a purely mechanical process, and one of his reviewers thought it "a model of logical reasoning." [47] Both Smith and his commentator were trapped by the same assumptions, and neither could have escaped from them except by admitting that there were some things in nature worth comparing with each other and some things that were not, and that there were no deductions that could legitimately be made before a lawful connection had been established. In other words, an analogy prior to experience could only be used to suggest an experiment, or, as the earlier theologians had suggested, to answer the a priori argument. To the American scientists of the second quarter of the nineteenth century, however, "sameness" was a before-the-fact judgment which could obviate experiment or even justify ignoring the results of experiment.[48]

The American scientists of the early nineteenth century were kept from making the required distinction for a number of reasons, two of which seem to have been the most important. First of all, they were convinced Baconians and, being such, had absolute faith in the inductive process and in the power of induction. At the same time they knew that their method had been subjected to severe attack and was in need of a sound logical foundation. Second, their characteristic holism, which took the form of a conviction that the world was an indivisible whole, created by one rational Being according to a plan—that is, their inability to conceive of closed systems or of irrelevancies—forced them to place all inductive generalizations on an equal footing. These two factors predisposed them toward falling easy prey to the first "unassailable basis" for science that they were offered. Wayland's "Philosophy of Analogy," grounded as it was in common-sense philosophy and appealing to the leading preconceptions of the time, was too seductive to resist. Had they begun making the required reservations for a proper use of analogies, it would have become necessary for them to rethink their entire philosophy, beginning with their fundamental assumptions about God's benevolence and His rational, unitary plan of creation. As long as their science rested upon such assumptions there remained a "science" of theology and of metaphysics from which their own research was, at bottom, hardly distinguishable.

That logical argument itself would have altered these convictions at this point is doubtful. What was required was a shock strong enough to shake the American scientific community from its comfortable mooring in the transatlantic version of Scottish common-sense philosophy. Two such shocks came shortly after mid-century, the effects of which were to break down the one exception scholars had always made to their belief in analogy—to deny the one "emergent" they had fought so hard to retain. First, the method of analogy produced thermodynamics, which seemed to deny both the world view of that generation and the distinction it had made between physics-chemistry and life. Then Darwin employed analogy to reach a conclusion that common sense rejected. Ironically enough, Von Helmholtz and Darwin were the legitimate methodological heirs of that scientific generation. Yet they were rejected by it, and between them, they finally destroyed it.

CHAPTER IX

෧෮෯

Science, Theology, and

Common Sense

Science, we like to think, is the highest product of man's reasoning powers—embodying the culmination of several thousand years' progress toward the ultimate in rationality and logic. As noted in the Introduction to this study, this strong presentist conviction presents a special danger to one who would characterize a period in the history of scientific thought. Put very bluntly, the historian is automatically tempted to employ a line of analysis that borders on moral judgment: one group of thinkers was "right," the rest "wrong"; one group followed the trail of scientific progress, as defined by our day; the rest were diversionary or even obscurantist. Commonly, such an analysis tends to develop as a dialectic between the "right" and the "wrong," with an accompanying tendency to oversimplify the amount of consensus on either "side," as well as the conflict between "sides." The general result, then, is a more or less false reading of the past.

To yield to this presentist temptation in trying to understand scientific thought in the early nineteenth century would be to do that period a special injustice. Broad areas of agreement and dispute indeed existed within the American scientific community, but it is not easily discernable that one segment was progressive and one was not. After this has been said, however, the historian writing from a twentieth-century advantage, if he is duly cautious, may define a certain developmental movement in American scientific thought during the thirty years from 1815 to 1845. Although hitherto no attempt has been made to display this movement to its best advantage, it can be traced through the preceding chapters. To view it as a comprehensive interpretation of the period would do violence to the historical record, but the development is worth outlining in some detail if only to highlight the ambiguities to be accommodated by any study of thinkers in a transitional intellectual era—and with not a single profound philosopher among them.

After the close of the War of 1812, it seemed to Americans that no more obstacles stood in the way of the rapid progress of scientific thought, with all the material and spiritual advantages it would supposedly bring with it. The real political independence achieved by the war was accompanied by a desire to extend that independence to economics, to literature, and to science. American scholars intensified their natural-history exploration in order to keep pace with the westward movement, developed a deep and abiding interest in research in the physical sciences, and began providing both an institutional basis for the pursuit of science and a domestic media for the dissemination of ideas and findings. Advance was so rapid that an almost childlike faith in science became the rule among educated Americans. The assumption underlying scientific work was that pure Baconianism—collection, description, and classification—if pursued long enough and consistently enough, would inevitably lead not only to a rich and mature understanding of nature, but also to great material happiness. And it was naturally assumed that such understanding and happiness would also promote a lofty morality and an unshakeable devotion to the Creator of all things. Benjamin Silliman's designation of his period as the "intellectual age of the world" is symptomatic of this first stage.

First becoming evident about 1820, however, a reaction to this gross optimism occurred. It was caused in part by the speculative excesses symbolized by "French materialism" and in part by a disillusionment with the failure of the Baconian method to lead automatically into the higher classifications that were its aim. The method had produced more unassimilated data than scientists could manage, and developments in the sciences had produced research interests to which the method seemed inapplicable. This explanatory bankruptcy of the Baconian philosophy was probably the most important factor in bringing about the second stage, which, as an unusually perceptive observer of the time noted, had two different manifestations.[1]

On the one hand, scholars offered wild and conflicting hypotheses in an abortive effort to account for the otherwise overwhelming data. This first manifestation was exemplified by John Esten Cooke's therapeutic system based on a single-entity theory of disease, by Lardner Vanuxem's two classes of matter out of which he thought a world could be deduced, by frantic and ill-founded efforts to achieve a "natural system" of classification, and by Robert Hare's decision that "matter" was merely an operationally defined term. This last, as we saw, led Hare finally into spiritualism. Such a bold disregard of Baconian precepts as manifested by Hare, Beck, and Vanuxem was not, however, as barren as it might appear at first glance, for it resulted in a deeper understanding of the nature of theory than had previously been held; whatever the occasional concomitant excesses, it did allow scientists for a while to work quite successfully with such imponderable entities as light, heat, and magnetism, that had only recently become of major scientific importance. James Maclurg stated as aptly as anyone—if a great deal more colorfully than most—the attitude of those who insisted that science must generalize even if it had to take new directions in order to do so. It was true, he said, that no system would be perfect until men could represent exactly the disposition of things as they flowed from the Almighty will; this meant that system would be eternally falling beneath the stroke of experience, for the best possible system could represent no more than the state of knowledge of the time in which it was formulated. Since this was true, it certainly followed that men should not be "wed to system," he said. But Maclurg did not think a marriage to system was necessary if one only desired to reap all the immediate benefits the system could offer; one should rather "treat it as a mistress," he said, drawing out the metaphor in greater detail, "embracing it with ardour at present, and discard it whenever we are disgusted with its defects, or attracted by the superior qualities of another." [2]

The other reaction was radically different. A large group of thinkers—probably the majority—was far more terrified by the theorizing than they were by the heaps of data the theories were supposed to explain. The frequent failure of hypotheses led them into the defeatist position that the "facts" were all of science. Science, it was argued by these thinkers, had nothing to do with the causes; it must rigorously confine itself to cross-sectional classification of a succession of events in time; neither could it have to do with the composition of bodies below the simple mechanical level, and most importantly of all it could explain none of the processes of vitality. It was alleged that the chemist, rather than simply investigating the processes of nature, was actually creating that which he studied, or at least changing it so much that it no longer pertained to nature. Explanation, according to this viewpoint, was to be in terms of the purposes of the Creator; it is important, however, to remember that these were not at first considered scientific explanations, but theological conclusions beyond the realm of science. If it did anything more than describe, science showed that God had purposes, as Frederick Beasley argued, but God's purposes, because they were multiple and indeterminate, could have nothing directly to do with science.

Some thinkers who displayed this positivistic tendency concerning science tried to rescue it from the inferior status to which it seemed to have been consigned by arguing that it had no need of the explanations it had been denied. Physiology could do quite well without the vital *principle*, as Samuel Jackson argued, as long as it could study vital *properties*, for these were all that really mattered. Lyell, in England, was the extreme representative in geology of this trend, but he had no prominent counterpart in America. Edward Hitchcock, in one of his periods at least, came as close as any eminent American to the extreme uniformitarian position of Lyell, but at least a mild form of catastrophism remained characteristic of American geologists.[3] Jackson and Hitchcock, however, could more aptly be described as precursors of the following generation than as representatives of their own, for there were not many among their contemporaries who followed their line of reasoning on these subjects. It was more often tacitly assumed that the vital principle would always preclude a real scientific understanding of physiological processes, and that geology could never do without a succession of catastrophes which could only be referred to the inscrutable will of the Creator. The question that James Maclurg was farsighted enough to raise—"But may not other Newtons, and other Franklins, extend still farther the boundaries of science?"—was not generally asked in America during the early nineteenth century.[4] On the whole, the finalistic attitude emerged with the discovery that the Humian argument could be used for the purposes of theology, however destructive it might be for science.

The third stage, beginning sometime in the late 1820s, was characterized by an effort to extend the inductive process through the doctrine of analogy. In many ways, analogy could be called the natural outcome of a clash between the two points of view just described. On the one hand, it was a way for scientists to deal with the unobservables that had caused such difficulty for the Baconian philosophy, and on the other hand, with its theological postulates and its teleological explanations, it could be acceptable to those who insisted that causality could only be described in terms of purpose. An analogy is only cross-sectional; it connects different elements within a temporal cross-section and implies nothing, in itself, about the cause. This is why it was as acceptable a mode of reasoning to Nathaniel Chapman, who insisted that science could not explain, as it was to Robert Hare, who was just as sure that science dealt in explanations. The doctrine of analogy, as developed by a line of thinkers from Wayland to Whelpley, was in one sense an importation into science of the final causes the positivists had wished to keep above science. In spite of this, however, and in spite of the excesses that often resulted, analogy would have to be called an advance to a better understanding of the nature of induction. It constituted an abandonment of the crudely additive view of induction that characterized the opening years of the century. Wayland's basic contention—that there was a great deal more to science than observation and a simplistic form of induction—was, after all, entirely correct. The movement of thought that occurred in the early nineteenth century could be said to have progressed from complete naivete to at least a modicum of sophistication in the use of the inductive method.

In an admittedly simplified fashion, the above outline does accurately represent an important line of development in scientific thought that occurred during the early nineteenth century. This pattern should not, however, be mistaken for a comprehensive interpretation of the meaning of the period. A number of considerations make this impossible. For example, Elisha Bartlett, who wrote his magnum opus at the very end of the period, had advanced further than anyone else of the time toward discounting the vital principle as a significant element in physiological thought. At the same time, his view of scientific method was so crudely additive that he was even forced to deny the legitimacy of induction.

There is another difficulty in that the same people keep turning up time after time in different contexts. Benjamin Silliman, for instance, was involved in every one of the stages, including the two rival reactions that Manley described. He thought in 1821 that he was living in the "intellectual age of the world" and that scientific progress would never cease. Yet, he was also one of those who tried to circumscribe severely the limits of science, and he fought unremittingly to preserve the vital principle and the uniqueness of life. At the same time, he was a follower of Hare in his chemical theories, and a collaborator with Hare in research; he accepted phrenology because the analogies favored it; and he eventually accepted the geological theory of a central fire even though he remained a strong defender of the veracity of Moses. If it were not for two embarrassing difficulties, one might, with a great deal of precedent, dismiss Silliman's ambiguities by calling him a "paradigm" of the movement of thought during that period. The first of these is that one would eventually have to conclude that there were almost as many "paradigms" as there were scientists, for, with the single exception of Elias Loomis, each of the people dealt with at length here, at one time or another, seemed to have held all of the positions described in this work. Hare, Beck, and Vanuxem spoke of the theoretical potentialities of science more often than of its limits, while Silliman and Olmsted spoke more often of its limits; but this is not to suggest that the first three never spoke of limits, nor the last two of potentialities. The differences were only in degree, resulting from the somewhat different interests of the scientists concerned. Richard Harlan wished to cast imagination quite out of science and "collect facts with philosphic circumspection," and he was willing—along with Jackson, Bartlett, and Chapman—to consign to the closet of the metaphysician any inquiries about ultimate causation. But he was also sure that the discovery of final causes was the great end of comparative anatomy, and it never occurred to him that there was any contradiction in the positions he was holding.

In the second place, referring again to Silliman for a ready example, there was no development over time in the positions he held. Indeed, within the space of a single year he had rebuked Horace Hayden for departing from Baconianism and suggested to Robert Hare that all of the imponderables might very well be different modifications of the same force.[5] Within the limits of one article,

John D. Godman denied that life was an emergent and suggested that one use the term "vital principle" to signify the "unknown modifying force" within the living organism. To cite just one more example, Lardner Vanuxem made contradictory statements about the nature of theory within the same year.

Indeed, if one expression were to be used to characterize the period as a whole, it would be "intellectual ambiguity"; it shared, not surprisingly, the central characteristic of the romantic era of which it was a part. This is not to say that men simply could not make up their minds about scientific issues, and that consequently they vacillated from position to position; on the contrary, it never seemed to occur to them that there was any necessity that they make up their minds about anything, for they saw no incompatibility. The assumptions of the time easily accommodated such equivocation, for the guiding ideas that were cherished were so flexible that nothing could rigorously follow from them. It is, accordingly, those ideas— or rather those complexes of ideas—that are of primary importance in any assessment of the period. Broadly speaking, they were three in number, and they have been appearing in one guise or another throughout this work.

The first was the idea of science as the great deliverer of mankind. Bacon had said that knowledge was power, and he had hoped to use science to elevate the condition of man beyond his wildest dreams and give him a control over nature which he had never known before. This was one Baconian idea rigorously held by all American scientists and, with the exception of disenchanted Catholics and High Church Episcopalians, by all those who commented on science. It made no difference whether one agreed with Vanuxem that science could provide a comprehensive understanding of the world, or with Chapman that it had certain rigidly prescribed limits beyond which it could not go; the dominant theme was still that science was the great hope for mankind, and that any progress that was made would be by virtue of the revolution begun by Francis Bacon. Vanuxem would have science perform its function by explaining the world to man; Chapman would leave the explanation to theology, but he still believed that man would control the physical world with the aid of science, insofar as he could control it at all, and elevate himself to new heights of health, longevity, and happiness.

Even among the intensely religious—among those who wanted above all to keep the idea of a personal God who was outside the causal order of science—the best that they could say for their Deity was that no theorem in mathematics was better proved than his existence. The best they could say for natural theology was that it was the *scientia scientiarum*. The fact that they rested their faith in God upon what they termed scientific evidence of his existence, and insisted that revelation would not be believable without anterior "scientific" proof of God's existence—the fact, that is to say, of the primacy of natural theology—demonstrates better than anything else the pervasiveness of the almost childlike faith in science that was a peculiar mark of that generation. Even intellectual disillusionment could not destroy that faith.

But along with this commitment to science—and here we find a major source of the ambiguity—there existed no really adequate idea of what science was, and there apparently existed no conception of the unity of science on any level below the theological first assumptions.[6] American thinkers did speak a great deal of the "scientific method" and they agreed almost to a man that science must be Baconian. Disillusionment with Baconianism brought only "reinterpretations" or "natural extensions"; it never brought acknowledged abandonment or even serious questioning. Lord Verulam was too firmly entrenched in America to be deposed during that generation, and only a few bold spirits ever suggested that classification was not the whole of science. Differences of opinion appeared only on the question of how far the classification would lead. But "Baconian" is a vague term, and I think the previous chapters have shown that almost anything could be done sincerely in the name of Bacon. One could readily agree with Tyler that "all science is classification" and still entertain contradictory statements about different "classifications." As long as unity was provided on the theological level, there was no need to seek it on the scientific. Neither did the generally held belief that there was "consistency of plan" throughout the whole creation demand any consistency in scientific statements; here again, the belief was framed in terms of purpose, and consistency was referred to the theological.

The second element was the Scottish philosophy of common sense. It is not too much to say that next to "Baconian philosophy," the most popular expression of early nineteenth-century intellectuals was "common sense." The highly touted and much overworked "faculties of the mind" and the naive realism of nineteenth-century Scottish philosophers, along with the psychology of John Locke "as interpreted and modified by Reid and Stewart," were made to order for a generation that had not yet learned to reconcile a devotion to science with a recognition that it could not give purpose and "meaning" to life. With a simplified version of the Scottish philosophy the scientists could find purpose in their minds and confuse it with purpose in external nature; they could find meaning in their own life and confuse it with a general "meaning in life"; and most importantly, they could find a need for understanding within themselves and reify their "understandings" in the world outside themselves. In a word, with the Scottish philosophy as a basis they could have both "science" and an anthropomorphic understanding of nature. This was the need that Scottish philosophy fulfilled, and it is no wonder that common sense was so highly, and so universally, praised. Not only did common sense provide ready, if specious, answers for all the problems of philosophy, but it also provided an ideal support for the third important element in the thought patterns of that period.

In a sense, the faculties of the mind and the much-appealed-to "composition" of human nature were taken as ultimate facts, but a statement like "man is so constituted that he cannot doubt the reality of whatever presents itself to his senses" is only a description, and it does indeed, as the natural theologians suggested, imply a Maker. This fact provided both need and support for the Protestant Christianity that lay at the very basis of all the thought of the period. The Protestantism of the early nineteenth century, finding its typical manifestation in natural theology, was "scientific" in the same sense that the popular Baconianism was scientific, and it was also as common-sensical as the Scottish philosophy. Baconianism provided a method of investigating nature, albeit a very crude and naive one, and the Scottish philosophy said that one must believe in the immediate sense impressions that were the raw materials of the Baconian philosophy. Justification for this belief, as well as justification for the belief that adding up particulars would some day yield a general, was provided by God as the natural theologians "deduced" his characteristics. Even more important, the existence of a benevolent Deity could be made to explain why the predictions made by science so often turned out to be true. And, like any ambiguous fundamental assumption, it could also explain why predictions sometimes failed. The failure of scientific predictions was an important part of the appeal to ignorance made by the finalists discussed earlier.

Science, theology, and common-sense philosophy woven together in an intricate pattern were the dominant themes in that period of American thought about the interpretation of nature. Given the combination of Baconian methodology and Scottish philosophy that was so evident both in England and in America, Samuel Tyler was perfectly correct in proclaiming that "philosophy, revelation, natural theology and physical science, are united in perfect harmony, proclaiming with one voice that there is a God." Had he added "and since this is so, our scientific method will inevitably lead us to truth," he would have been making a circular argument, but at the same time he would have summarized the thought of the period in a single sentence.

APPENDIX I

∾

Biographical and Bibliographical

Sketches of Fifty-five

Leading American Scientists of

the Period 1815 to 1845

The men covered in this appendix were the most prolific contributors to the sixteen scientific journals which were indexed for this study. As such, they form the nucleus about which this study is built. One other man should appear in the list, but he flourished only briefly and I could find no information about him; not even so much as an obituary notice.[*]

The relative lengths of the sketches should not be considered as indicative of a judgment of the scientist's importance. Instead, I have tried to give more complete information concerning less well-known figures than those that are already better known. There is very little, for example, that one could add to scholarly knowledge of James D. Dana in a sketch of one or two pages; for some others, like John D. Godman, a great deal of generally unknown information can be presented. I have not duplicated information found in the text.

The same considerations dictated the number of bibliographical entries allotted to a subject; generally speaking, where a major scholarly biography has been written, I have listed only one or two contemporary memoirs, if any were written, in addition to the biography. On the other hand, where no major biography has been written, I have listed all the sources which have come to my attention, except those that were obviously based upon previous works and add nothing new about the man.

[*] A. B. Quinby, who published articles on mechanical power conversions and steam engines in the *American Journal of Science and Arts* between 1824 and 1827, is the man whose sketch is missing. The only information I have about him is that he lived in New York at the time he was publishing. There is no record that he ever wrote a book, and I found no mention of him in any other journals of the period.

Similarly, I have only listed manuscript sources which, to my knowledge, have not been used by scholars. In some cases, however, brevity does indicate a poverty of source material, rather than adequate coverage elsewhere.

The following abbreviations or short forms are used throughout for works frequently referred to:

AAAS *Proc.*	American Association for the Advancement of Science, *Proceedings*
AJS	*American Journal of Science and Arts*
APS, *Proc.*	American Philosophical Society, *Proceedings*
Bell	Whitfield Bell, *Early American Science: Needs and Opportunities for Study*
Drake's *DAB*	Francis S. Drake, *Dictionary of American Biography, Including Men of the Time.* . . . (Boston, James R. Osgood and Co., 1876).
CAB	*Appleton's Cyclopedia of American Biography*
CAMB	Howard A. Kelly, *A Cyclopedia of American Medical Biography, Comprising the Lives of Eminent Deceased Physicians and Surgeons from 1610 to 1910* (Philadelphia and London: W. B. Saunders Company, 1912), 2 Vols.
DAB	*Dictionary of American Biography*
Gross, *Autobiog.*	*Autobiography of Samuel D. Gross, With Sketches of His Contemporaries,* 2 Vols. (Philadelphia, 1861).
Gross' *Lives*	Samuel D. Gross, *Lives of Eminent American Physicians and Surgeons of the Nineteenth Century* (Philadelphia, 1861).
NAS *Biog. Mem.*	*National Academy of Science, Biographical Memoirs.*
Proc. AAA&S	*Proceedings of the American Academy of Arts and Sciences*
Simpson	Henry Simpson, *Lives of Eminent Philadelphians Now Deceased* (Philadelphia, 1859).
Youmans	William J. Youmans, *Pioneers of Science in America* (New York: D. Appleton and Company, 1896).

For those works which are in dictionary form I have not given page numbers.

ALEXANDER DALLAS BACHE (1806–1867), the great-grandson of Benjamin Franklin, was born in Philadelphia, and after attending a classical school graduated from West Point with highest honors at the age of 19. He then taught for three years, served briefly as a Lieutenant of Engineers, and in 1828 became professor of natural philosophy and chemistry at the University of Pennsylvania. Upon being appointed head of the newly founded Girard College, he went abroad to study educational institutions, and upon his return published *Education in Europe,* which is said to have done much to improve educational methods in the United States. During the period 1829–1837 he published articles in astronomy, meteorology, chemistry, and physics, more than half of which were in astronomy. He had early developed an interest in terrestrial magnetism, and largely through his efforts the first magnetic observatory in the

United States was established at Girard College in 1840. In 1843 he succeeded Hassler as superintendent of the United States Coast Survey, a position which he held until his death. Bache was a member of the American Philosophical Society, the Royal Society of London, was one of the incorporators of the Smithsonian Institution, and served as the first president of the National Academy of Science, which he, along with other members of the scientific Lazzaroni, had been influential in founding. He also served terms as president of the American Philosophical Society and the American Association for the Advancement of Science. For biography, see B. A. Gould's memoir in AAAS *Proc.*, 1869, pp. 1–47; Joseph Henry's Memoir in NAS *Biog. Mem.*, I (1877), 183 ff., both of which contain a complete bibliography; or Merle M. Odgers, *Alexander Dallas Bache, Scientist and Educator* (Philadelphia, 1947). Nathan Reingold is currently working on a study of Bache which should add considerably to our knowledge of the scientific Lazzaroni, an elitist group of American scientists of which he was a key member.

JOHN BACHMAN (1790–1874) was born in Dutchess County, New York, the son of a well-to-do farmer. He attended Williams College, and in 1815, after an attack of tuberculosis, moved to Charleston, South Carolina, where be became minister of St. John's Lutheran Church. Bachman, a member of the American Academy of Arts and Sciences, was a leader in founding the South Carolina State Horticultural Society in 1833. He received a Ph.D. from the University of Berlin in 1838. All of Bachman's publications, 1834–1842, were in descriptive zoology and ornithology. In collaboration with Audubon, he published a three-volume work, *The Viviparous Quadrupeds of North America*, 1845–1849. For his later, controversial work on the unity of the human species, see William M. Stanton, *The Leopard's Spots: Scientific Attitudes Toward Race in America, 1815–1859* (Chicago, 1960). For biography, see J. B. Haskell and C. Bachman, *John Bachman, the Pastor of St. John's Lutheran Church, Charleston, S.C.: Letters and Memoirs of His Life* (Charleston, 1888); John F. Ficken, *A Sketch of the Life and Labors of John Bachman* (Charleston, 1924); or Donald C. Peattie's article in *DAB*. The Bachman papers are in the Charleston Museum.

JACOB W. BAILEY (1811–1857) was born in Ward (now Auburn), Massachusetts, the son of a minister. After attending the common school in Providence, Rhode Island, he graduated fifth in his class at West Point in 1832. From 1834 to 1857 he served as professor of chemistry, mineralogy, and geology at West Point. Although he published work in chemistry, botany, mineralogy, geology, physiology, and meteorology, 1837–1844, his chief interest was in the new field of micro-paleontology, in which he pioneered in this country. For biography, see Stanley Coulter's article in the *Botanical Gazette*, XIII (1888), 118–24; A. A. Gould's memorial in AAAS *Proc.*, 1857–1858, pp. 1–8, or the article by Coulter in *DAB*.

Two volumes of his bound manuscript letters are in the Boston Society of Natural History, and the Huntington Library has letters to Bache which contain accounts of his analysis of deep-sea soundings.

LEWIS C. BECK (1798–1853) was born in Schenectady, New York, a member of a prominent scientific family. After graduating from Union College, he studied medicine under Doctor Thomas Dunlop and received an M.D. from the College of Physicians and Surgeons of the University of New York. His journal publications, 1826–1844, were in chemistry, botany, mineralogy, geology, physiology, and meteorology. More than half were in chemistry, which was his field of special competence. He held various academic posts, including that of professor of botany at the Rensselaer Polytechnic Institute (1824–1829), professor of botany at the Vermont Academy of Medicine (1826–1832), professor of chemistry and natural history at Rutgers (1830–1837) and professor of chemistry and pharmacy at the Albany Medical College (1841–1853). In 1837 he served as mineralogist to the geological survey of New York. For biography, see *CAB*, 5:442; Lyman F. Kebler's article in *DAB*; sketches by J. V. P. Quackenbush in *Transactions of the New York Medical Society*, 1854, p. 63; Mrs. C. E. Van Cortlandt in *Annals of the Medical Society of the County of Albany*, 1864. For an evaluation of his work as an analytical chemist, see Kebler's "A Pioneer in Pure Food and Drugs: Lewis C. Beck," in *Industrial and Engineering Chemistry*, September 11, 1924. Alden March in Gross' *Lives*, pp. 679–96, has compiled an excellent bibliography, and relates some information which is apparently based upon personal knowledge of the subject.

CHARLES L. BONAPARTE (1803–1857), Prince of Canino and Musignano, and the nephew of Napoleon, was born in Paris. During the revolution of 1848 he was one of the leaders of the Republican Party at Rome, and in 1849 was Vice President of the Constituent Assembly. During his residence in the United States (1825?–1830?), he published articles on ornithology in the journals and a four-volume continuation of Wilson's *Ornithology*. He was a member of the American Academy of Arts and Sciences. Bonaparte's principal work was *Iconographia della Fauna Italica* in three volumes (Rome, 1835–1845). For biography, see Drake's *DAB* or *Memoirs of Charles L. Bonaparte*, written by himself (New York, 1836).

TIMOTHY A. CONRAD (1803–1877) was born in Trenton, New Jersey, the son of a publisher and printer. He received his early training at Friends' schools, and in 1829 was appointed professor of botany at the University of Pennsylvania. In 1837 he was appointed New York State Geologist. His publications, 1840–1842, were in geology, paleontology, and zoology. He was one of the early geologists in America to argue for dating of geological strata by organic remains (*AJS*, XXVIII, 1835, 104,

280), but he remained a devoted follower of Agassiz against Darwin to the last. In addition to his scientific work, Conrad wrote a great deal of poetry, much of which was published in the Philadelphia newspapers. He was a member of the American Philosophical Society and the Academy of Natural Sciences of Philadelphia. For biography, see *CAB*, 8:466; or Youmans, pp. 385-93. For lists of his papers, see *Miscellaneous Publications of the United States Geological Survey of the Territories*, N. 103, and C. A. White and H. A. Nicholson, *Bibliography of the North American Invertebrate Paleontology* (Washington, 1878).

JAMES DWIGHT DANA (1813–1895) was born in Utica, New York of wealthy parents. He entered Yale in 1830 but left before graduation in 1833. From 1838 to 1842 he was on an around-the-world cruise as mineralogist and geologist for the Wilkes' Expedition. From 1850 to 1895 Dana was professor of natural history (later professor of geology and mineralogy) at Yale. His publications, 1835–1844, were in mineralogy, zoology, and geology. Dana, who married Benjamin Silliman's daughter, became an editor of the *American Journal of Science and Arts* in 1840, and after Silliman's retirement from active editorship he made the journal virtually an organ for the scientific Lazzaroni. Dana was a member of the Royal Society of London, President of the American Academy of Arts and Science, member of the American Philosophical Society, and one of the original members of the National Academy of Science. For biography, see Daniel C. Gilman, *The Life of James Dwight Dana* (New York, 1899). Yale has manuscripts, letters, etc., and there is some scattered correspondence in the Huntington Library and the Historical Society of Pennsylvania.

JAMES FREEMAN DANA (1793–1827) was born in Amherst, New Hampshire, the son of a Naval officer. He attended Exeter Academy, was graduated from Harvard in 1813, and in 1815 studied in England. After receiving an M.D. in 1817, he became professor of chemistry at the University of New York. His publications, 1818–1828, were in chemistry, with one article in geology. In 1818, with his brother, Samuel L. Dana, he published *Outlines of the Mineralogy and Geology of Boston*, the first description of the geology of the area. For biography, see *CAB*, 9:402, or Lyman C. Newell's article in *DAB*. *The American Chemist*, V (1874–1875), Nos. 2, 3, 6 has a short account of his work.

JAMES E. DEKAY (1792–1851) was born in Lisbon, Portugal, the son of a British Army Captain. He attended school in Connecticut and received a degree in medicine from Edinburgh. During most of his life he was a practicing physician. In addition to his journal publications, 1822–1837, in paleontology, zoology, and geology, he published *Zoology of New York* (1842–1844), 6 vols., as part of the state survey. For biography, see

CAB, 9:402; William Shaw's article in *DAB or* H. L. Fairchild, *History of the New York Academy of Science* (New York, 1887). The Library of Congress has a typewritten copy of a manuscript prepared by DeKay entitled "The Book of the Children of DeKay."

CHESTER DEWEY (1784–1867) was born in Sheffield, Massachusetts and after receiving a common-school education was graduated from Williams College in 1808. He was ordained a Congregational minister in 1810, received an honorary M.D. from Yale in 1825, a D.D. from Union College in 1838, and an LL.D. from Williams College in 1850. Dewey was professor of mathematics and natural philosophy at Williams College (1810–1827), professor of botany and chemistry in the Medical College at Pittsfield, Massachusetts (1827–1836), principal of the Collegiate Institute at Rochester, New York (1836–1850), and professor of Chemistry and natural history at the University of Rochester (1850–1860). He published (1818–1843) articles in botany, geology, mineralogy, meteorology, and chemistry. He was best known for his extensive studies of the Carices published in the *American Journal of Science* from 1824 to 1840. Dewey was a member of the American Academy of Arts and Sciences and the American Philosophical Society. For biography, see Charles W. Dodge's article in *DAB*, the obituary by Asa Gray in *AJS*, XLV (1868), 122–23; Martin B. Anderson's sketch in *Smithsonian Reports* for 1870; or Charles W. Seely's sketch in *Proceedings of the Rochester Academy of Science*, III (1891). The Rochester Historical Society has Dewey family records, including a manuscript four-page sketch of Chester Dewey by his son.

DANIEL DRAKE (1785–1852) was born of poor and uneducated parents in Plainfield, New Jersey. After a few months education in a country school he was apprenticed to Dr. Goforth, a well-known physician in Cincinnati. In 1816 he received an M.D. from the University of Pennsylvania. For the remainder of his life he was a professor at medical schools, chiefly at Transylvania, but also at the Jefferson Medical College of Philadelphia, the Cincinnati Medical College, the Medical College of Ohio, and the University of Louisville. He also edited a number of journals, including the *Western Journal of the Medical and Physical Sciences*. His chief work was *A Treatise on the Principal Diseases of the Interior Valley of North America* (1850), a pioneering work on ethno-medicine. During the period 1825–1830, he published articles in physiology, philosophy, and geology. Autobiographical material is in Daniel Drake, *Pioneer Life in Kentucky* (Cincinnati, 1870), edited by his son. For biography, see D. Juettner, *Daniel Drake and His Followers* (Cincinnati, 1909); or Emmet F. Horine, *Daniel Drake, 1785–1852, Pioneer Physician of the Midwest* (Philadelphia: University of Pennsylvania, 1961).

AMOS EATON (1776–1842) was born in Chatham, New York and graduated from Williams College in 1799. He was a practicing lawyer for a few years, but ended this career after being sentenced to prison for life on charges which have not yet been satisfactorily cleared up. He was pardoned by Governor De Witt Clinton, and in 1815 began a two-year period of study with Silliman at Yale. In 1817 he began his courses of lectures on natural science, and in 1818, at Governor Clinton's invitation, he delivered a course of lectures on chemistry and geology before the members of the New York legislature. In 1824 he became senior professor at Rensselaer Institute. During the period 1818–1840 he published articles on geology, chemistry, philosophy, mineralogy, and surveying. For biography, see Calvin Durfee, *A Sketch of the Life and Services of the Late Professor Amos Eaton* (Boston, 1860); E. M. McAllister, *Amos Eaton, Scientist and Educator* (Philadelphia, 1941); George P. Merrill's article in *DAB;* or Palmer C. Ricketts, *Amos Eaton, Author, Teacher, Investigator . . . (Rensselaer Polytechnic Institute, Engineering and Science Series, No. 45)* (November, 1933).

JOHN PATTEN EMMET (1796–1842) was born in Dublin, Ireland, the son of the well-known Irish nationalist lawyer and orator, Thomas Addis Emmet. He was brought to the United States at an early age and received all his schooling in this country. He was educated at West Point, but ill health prevented his graduation, and he went to Italy for a year. Afterwards, he returned to New York, taking a degree at the College of Physicians and Surgeons in 1822. One of Thomas Jefferson's original appointees at the University of Virginia, he was professor of natural history from 1825 to 1827 and professor of chemistry and materia medica from 1827 to 1842. He published, 1830–1837, articles in chemistry, mostly having to do with galvanic electricity. Emmet was also a frequent contributor to the *Virginia Literary Museum.* He was a member of the American Philosophical Society. For biography, see T. A. Emmet, *Biographical Sketch of Professor John Patten Emmet, M.D., and Other Members of the Emmet Family* (New York, 1898); Robert M. Slaughter's article in *DAB;* the sketch in *The Alumni Bulletin,* University of Virginia, vol. 1, No. 4 (February, 1895); or George Tucker, *Memoir of John Patten Emmet* (Philadelphia, 1845). The Emmet papers are at the University of Virginia.

JOSIAH W. GIBBS (1790–1861) was born in Salem, Massachusetts of a substantial family. He graduated from Yale in 1809 and returned as a tutor from 1811 to 1815, after studying Hebrew and other Oriental languages at Andover with Moses Stuart. He was licensed to preach in 1814. In 1824 he became lecturer in Biblical literature at Yale, and from 1826 until his death in 1861 served as professor of sacred literature at the Yale Divinity School. The leading philologist of his day, he published articles on that subject from 1833 to 1843. For biography, see Charles C. Torrey's article in *DAB;* F. B. Dexter, *Biographical Sketches of Graduates of Yale College, 1805–1815* (1912), with a complete bibliography; G. P. Fisher, "Discourse Commemorative of Josiah Willard Gibbs, LL.D.," in the *New*

Englander for July, 1861; Timothy Dwight, *Memoirs of Yale Life and Men* (New York, 1903), pp. 265–77; the article by George E. Day in W. L. Kingsley, *Yale College* (New York, 1879), II, pp. 37–40; or J. W. Gibbs, *Memoir of the Gibbs Family* (Philadelphia, 1879), *passim.*

JOHN D. GODMAN (1794–1830) was born in Annapolis, Maryland and apprenticed to a printer at the age of seven. He worked as a printer until 1814, after which time he began the study of medicine, receiving his degree from the University of Maryland in 1818. In 1819 Daniel Drake appointed him professor of surgery and obstetrics at the Ohio Medical College, but he left after the 1820 controversy among the professors of the college. Godman was professor of anatomy at Rutgers from 1824 to 1830. He edited the first medical journal west of the Allegheny Mountains, the *Western Quarterly Reporter of the Medical and Physical Sciences;* and was later an editor of the *Philadelphia Journal of Medical Science,* and the *Journal of Foreign Medical Science and Literature.* Godman is remembered for his *American Natural History* (1826–1828), his edition of *Bell's Anatomy and Physiology* (1827), and his *Anatomical Investigations* (1824). During his short life he taught himself to read Latin, Greek, German, French, Italian, and Dutch, using this knowledge to translate articles for his journals. Godman was a member of the American Philosophical Society. For biography, see T. Sewall, *Memoir of Dr. Godman* (New York, 1837); William S. Miller's article in *Annals of Medical History,* IX (1937), 293–303; Daniel Drake's memoir in the *Western Journal of Medical and Physical Science,* January, Feburary, March, 1831; Gross' *Lives,* pp. 247–66; Gross' *Autobiog.,* pp. 44–46; E. F. Cordell, *The Medical Annals of Maryland* (Baltimore, 1903); P. S. Godman, *Some Account of the Family of Godman* (London, 1897). To the third edition of *American Natural History* there was prefaced a biographical sketch of the author, vol. I. pp. iii-xxvi.

ASA GRAY (1810–1888), one of America's few early scientists of international repute, was born in Sauquoit, New York, the son of a tanner. After attending courses of lectures at the College of Physicians and Surgeons of the Western District of the State of New York (Fairfield), he received an M.D. degree in 1831 and began to practice medicine. Gray is remembered in the history of science chiefly for his studies of the distribution of plants, which Darwin used in developing his theory of natural selection, and for his early attempt to accommodate Darwinian evolution to religion. In 1842 Gray became Fisher Professor of Natural History at Harvard, and he remained at that institution until his retirement in 1873. For an excellent biography of Gray and an analysis of his contributions to botany, see A. Hunter Dupree, *Asa Gray* (Cambridge, 1959).

JACOB GREEN (1790–1841) was born in Philadelphia and graduated from the University of Pennsylvania in 1806, at which time he was licensed to practice law. He was the son of the Reverend Ashbel Green, president of Princeton. From 1818 to 1822 he was professor of chemistry, experimental philosophy, and natural history at Princeton; and from 1822 to 1841 professor of chemistry at Jefferson Medical College. His publications, 1818–1839, were in paleontology, chemistry, zoology, and anthropology. Green was best known for his compilations in the field of chemistry; he published several textbooks, following Turner. His best-known original contribution was in the field of paleontology, *Monograph of the Trilobites of North America* (1832), again largely a compilation, but embodying a great deal of his own research. His Princeton lectures were published in 1827 as *Electro-Magnetism*. He received an honorary M. D. from Yale in 1827. For biography, see *CAB*, 13:552; Edgar F. Smith, *Jacob Green, 1790–1841, Chemist* (Philadelphia, 1923); Horace B. Baker's article in *DAB;* S. S. Greene, *A Genealogical Sketch of the Descendents of Thomas Green(e) of Malden, Massachusetts* (Providence, 1858), or Simpson.

SAMUEL STEDMAN HALDEMAN (1812–1880) was born in Lancaster County, Pennsylvania, of a substantial family. He studied under H. D. Rogers at Dickinson College from 1828 to 1830, but did not graduate. In 1837 he worked as an assistant to H. D. Rogers on the Pennsylvania Geological Survey. From 1842 to 1843 he delivered a course of lectures on zoology at the Franklin Institute; from 1851 to 1855 was professor of natural history at the University of Pennsylvania; from 1853 to 1858 professor of geology and chemistry at the Pennsylvania Agricultural College, and professor of natural sciences at Delaware College. His publications, 1839–1844 were in zoology and chemistry. In an article in the *Boston Journal of Natural History*, IV (1844), 468 ff., he suggested a need for a reformulation of the species concept, and observed that Lamarck had been dismissed too readily from serious consideration. In 1844 he became a Catholic, and in 1847 he made his first trip to Europe. After the late 1840s he devoted himself entirely to philological studies, becoming in 1868 the first professor of comparative philology at the University of Pennsylvania and holding this position for the remainder of his life. Haldeman was a member of the American Philosophical Society and the American Academy of Arts and Sciences. For biography, see J. P. Lesley's memoir in NAS *Biog. Mem.*, II (1886), 141 ff.; George C. Harvey's article in *DAB;* John Livingston, *Portraits of Eminent Americans Now Living*, vol. IV (1854); Alexander Harris, *Biographical History of Lancaster County, Pennsylvania*, (Lancaster, 1872); Memoir by D. B. Brinton in *Proc.* APS, XIX (1880–1881), 279–85; C. H. Hart's memoir with bibliography by Mrs. Eliza Figyelmesy, his daughter, in *Pennsylvania Monthly*, August, 1881; H. L. Haldeman (nephew), memoir and bibliography in *Records of the American Catholic Historical Society*, September, 1898; and P. C. Croll's article in *The Pennsylvania German*, July, 1905. Haldeman letters are in the Academy of Natural Sciences of Philadelphia and the Historical Society of Pennsylvania.

ROBERT HARE (1781–1858) was born in Philadelphia, the son of a wealthy brewer and state senator. He attended school in Philadelphia and received an honorary M.D. from Yale in 1806 and from Harvard in 1816. He spent his academic life (1818–1847) as a professor of chemistry in the Medical School of the University of Pennsylvania. Over the twenty-five year period from 1818 to 1843 he published 119 articles in chemistry and a number of articles in meteorology. He first attracted the attention of the scientific world at the age of twenty, when he invented the oxy-hydrogen blowpipe, which proved to be the source of the highest degree of heat then known. Known as a brilliant experimenter, with a flare for the colorful and a penchant for large-scale demonstrations, he was constantly inventing new apparatus for his chemistry classes. In 1819 he invented the calorimotor, a galvanic apparatus which became the model for Plante's secondary battery, and in 1821 the deflagrator for generating high electric currents. In 1839 he received the Rumford Medal from the American Academy of Arts and Sciences. During his lifetime Hare carried on active controversies with the most important scientists of the nineteenth century, including Berzelius over chemical nomenclature, Faraday over electricity and the nature of matter, William Whewell over the existence of imponderables (Hare was a firm believer), W. C. Redfield over the theory of storms, and with all comers on the issue of the materiality of heat, a view which he never relinquished. In 1854 he became converted to spiritualism. Hare was a member of the American Philosophical Society and the American Academy of Arts and Sciences. For biography, see Edgar F. Smith, *The Life of Robert Hare* (Philadelphia, 1917); B. Silliman in *AJS*, July, 1858, pp. 100–105; G. P. Fisher, *Life of Benjamin Silliman* (1866), I, 98 ff.; Gross, *Autobiog.*, II, 297–98; Simpson; J. W. Jordan, *Colonial and Revolutionary Families of Philadelphia* (Philadelphia, 1911), I, 129–31; Youmans, pp. 176–81. The Hare papers are at the University of Pennsylvania.

RICHARD HARLAN (1796–1843) was born in Philadelphia and after studying under Joseph Parrish took a medical degree at the University of Pennsylvania in 1818. He then began the practice of medicine in the city of his birth. In 1821 he became professor of comparative anatomy at the Philadelphia Museum. In 1838 he visited Europe, and after his return settled in New Orleans, where he spent the remainder of his life. Over half of his journal publications were in descriptive zoology, a large number were in vertebrate paleontology, and the remaining few were in physiology, chemistry, and anthropology. In 1825 he published *Fauna Americana*, his most notable work and the first systematic treatise on American mammals by an American. This was largely a compilation, based upon A. G. Desmarest's *Mammalogie* (1820–1821), although some of his own work was incorporated in it. In the work, he grouped fossil forms with what he presumed to be their nearest living representative, and became the first American to apply Linnaean names to American fossil vertebrates. For this work he received so much hostile criticism that planned subsequent volumes were never published; on this point, see his *Refutation of Certain Misrepresentations Issued Against the Author of the Fauna Americana* (1826). In 1835 he published a volume of his most important essays, *Medical and Physical Researches: Or Original Memoirs in Medicine, Surgery, Physiology, Geology, Zoology, and Comparative Anatomy* (Philadelphia, 1835). Harlan was a member of the American Philosophical Society. For biography, see A. H. Harlan, *History and Genealogy of the Harlan Family* (1914); Simpson, pp. 247–66; Daniel Fisk's article in *DAB*, or a brief account in *CAMB*. For Harlan's place in the history of paleontology, see George G. Simpson, *The Beginnings of Vertebrate Paleontology in North America, Proc.* APS, LXXXVI (1943), 161–64.

Augustus A. Hayes (1806–1882) was born in Windsor, Vermont, and was graduated from Captain Partridge's military academy at Norwich, Vermont, in 1823. He later studied chemistry under J. F. Dana and was an assistant professor of chemistry at New Hampshire Medical College. In 1828 he moved to Boston to become a consulting chemist to dyeing and bleaching companies, and iron and copper smelting establishments. He directed a color manufactory at Roxbury. His publications, 1828–1844, were predominantly in chemistry, with the exception of two meteorological articles in 1830. In 1846 he received an honorary M.D. from Dartmouth. Out of a series of investigations begun in 1825 on the proximate principles of American medicinal plants, he discovered the vegetable alkaloid, sanguinarine, found in the blood root. His research in 1838 upon the relative value of fuels led to important improvements in steam boilers; and he was the first to suggest the application of oxides of iron in the refining of pig-iron. He also devised methods of shortening the time needed in refining copper and in smelting iron. Hayes was a member of the American Academy of Arts and Sciences. For biography, see *CAB*, 11:56; Lyman C. Newell's article in *DAB;* W. A. Ellis, *Norwich University, 1819–1911* (Montpelier, Vermont, 1911); *The Boston Transcript* for June 23, 1882; a memoir in *Proc. AAA&S*, n.s., X (1883); and J. J. Dana, *Memoranda of Some of the Descendents of Richard Dana* (Boston, 1865), for an account of his work with J. F. Dana.

Joseph Henry (1797–1878) was born in Albany, New York, where he attended the common school until the age of 14, when he was apprenticed to a jeweler. He later studied at the Albany Academy, where he tutored the S. Van Rensselaer family and assisted Beck in chemistry. In 1826 he became professor of mathematics at the Albany Academy. In 1837 he spent nine months studying in Europe, and later became professor of philosophy at Princeton. In 1846 he became the first secretary of the Smithsonian Institution and held the position for the remainder of his life. His publications, 1831–1840, were all in physics-chemistry, primarily concerned with electromagnetism. In 1830, independently of Faraday, he discovered electromagnetic induction, and his work made possible the invention of the telegraph by Samuel F. B. Morse. His only scientific book, *Syllabus of Lectures on Physics,* was published in 1844 while he was teaching at Princeton. Henry was a member of the American Philosophical Society, the American Academy of Arts and Sciences, the National Academy of Science, and was an important member of the informal group known as the scientific Lazzaroni. For biography, see Thomas Coulson, *Joseph Henry* (Princeton, 1950); *A Memorial of Joseph Henry,* published by order of Congress in 1880 (included also in Smithsonian Miscellaneous Collections, Vol. XXI), which contains several biographical sketches and a complete bibliography. For an account of his part in the discovery of electromagnetic induction, see Mary A. Henry, "America's Part in the Discovery of Magneto-Electricity—A Study of the Work of Faraday and Henry," *Electrical Engineer* (New York), January 13–March 9, 1892. His *Scientific Writings* were published by the Smithsonian Institution as *Smithsonian Miscellaneous Collections,* Vol. XXX. For a personal account of his work on electrical induction, see his "Memoirs," in J. S. Ames (editor), *Discovery of Induced Electric Currents* (New York, 1900). Henry's lectures at Princeton, compiled from his syllabus, laboratory journal, and student notes, are now being edited by Charles Weiner of the American Institute of Physics, and Nathan Reingold of the Smithsonian Institution has begun a projected twenty-volume edition of Henry's papers.

N. M. HENTZ (1797–1856) was born in Versailles, France, and studied medicine at the Hospital Val de Grâce until the fall of Napoleon, when his father fled to America. From 1824 to 1830 he was professor of modern languages at the University of North Carolina. From 1830 to the end of his life, he conducted girls' schools, including schools in Covington, Kentucky (1830–1832), Cincinnati (1832–1834); Florence, Alabama (1845–1848); and Columbus, Georgia (1848–1849). His publications, (1821–1844) were all in descriptive zoology, mostly on spiders, about which he became the leading authority in America. Hentz' "Monography of the Spiders of the United States," published in the *Boston Journal of Natural History* between January, 1842, and December, 1847, is still consulted by entomologists. For biography, see *CAB,* 1:428, or *DAB.*

EDWARD C. HERRICK (1811–1862) was born in New Haven, Connecticut, the son of a noted teacher, from whom he received a classical education. He was first in business as a bookseller, and in 1843 became the librarian of Yale College. His publications, 1838–1842, were all in astronomy with a single exception, an article in descriptive zoology. He was the discoverer of the returning August meteoric showers, although Quetelet's similar theory predated his. In 1835 he received an honorary M. A.

degree from Yale College. For biography, see *CAB*, 11:170; the article by Harris E. Starr in *DAB;* or William L. Kingsley, *Yale College, A Sketch of Its History,* Vol. I (1879).

SAMUEL P. HILDRETH (1783–1863) was born in Methuen, Massachusetts, the son of an eminent surgeon. He first attended Phillips, Andover, and Franklin academies; later beginning the study of medicine with his father, Dr. Samuel Hildreth, and Dr. Kittredge of Andover. He attended a course of lectures at Cambridge University. In 1805 he received a degree in medicine from the Massachusetts Medical School. Hildreth was in the New Hampshire legislature in 1810–1811, during which time he secured passage of a law regulating the practice of medicine and providing for medical societies. For most of his life he practiced medicine in Marietta, Ohio. His publications, 1826–1844, were mostly composed of minute meteorological tables and annual synopses, but he did some work in geology, paleontology, geography, and zoology, all primarily descriptive. The Hildreth papers are in Marietta College. For biography, see the article by Henry E. Henderson in *CAMB;* Albert P. Mathews in *DAB;* Hildreth's own *Genealogical and Biographical Sketches of the Hildreth Family* (Marietta, Ohio) published in 1840; or the sketch by John Eaton in *Memorial Biographies of the New England Historical and Genealogical Society.* His meteorological observations were compiled and discussed by C. A. Schatt in *Smithsonian Contributions to Knowledge,* 1870.

EDWARD HITCHCOCK (1793–1864) was born in Deerfield, Massachusetts, the son of a hatter. In 1820 he was graduated from the Yale Theological Seminary. He studied chemistry and geology with Silliman, and from 1821 to 1864 was professor of chemistry and natural history at Amherst. He also served as president of Amherst from 1845 to 1854. Hitchcock was head of the geological survey of Massachusetts from 1830 to 1833. His degrees include: A.M., Yale, 1818; LL.D., Harvard, 1840; DD., Middlebury, 1846. He was a member of the American Philosophical Society and the American Academy of Arts and Sciences. In 1860 he became the first president of the American Association of Geologists, and in 1863 he was one of the original members of the National Academy of Science. In 1850 he was commissioned by the state of Massachusetts to examine agricultural schools in Europe. Almost half of his publications, 1818–1844, were in geology; the remainder were scattered throughout the range of sciences. Hitchcock is best known for his studies of what he took to be fossil bird footprints (now known to have been dinosaur tracks) beginning in 1836. Hitchcock developed an interest in natural theology later in life, publishing in 1851 *The Religion of Geology and Its Connected Sciences.* For biography, see his *Reminiscences of Amherst College,* containing autobiographical material on the most productive years of his life; the sketch by J. P. Lesley in NAS *Biog. Mem.* I (1877),

115; the article by George P. Merrill in *DAB;* C. H. Hitchcock's memoir in *American Geologist,* XVI (1895), 133–49, with an excellent bibliography; or Youmans, pp. 290–99.

CHARLES THOMAS JACKSON (1805–1880) was born into a substantial family in Plymouth, Massachusetts. He had medical training under James Jackson and Walter Channing before graduating from Harvard Medical School in 1829; afterward he studied in Paris and Vienna. While at Vienna he assisted in dissecting 200 victims of the cholera epidemic. In Paris he became acquainted with the statistical methods then being introduced into medicine by Pierre Louis and his empirical school. After returning to the United States in 1836, he opened a laboratory in Boston for instruction in analytical chemistry—the first laboratory of its kind in America to receive students. Jackson has a number of other "firsts" on his record: he was among the first in America to engage in agricultural chemistry, the first to make a practical application of the telegraph, and the first to demonstrate the anesthetic properties of ether. In these last two, he engaged in long controversies over priority with those who developed his ideas; similarly, after C. F. Schonbein announced the discovery of guncotton in 1846, Jackson claimed priority. He was engaged in controversy most of his life until he lost his mind in 1873. His publications, 1828–1844, were in geology, chemistry, and mineralogy. In addition to operating his chemistry laboratory, he was geologist on the Maine geological survey, 1837–1839; the Rhode Island survey, 1840; and the New Hampshire survey, 1841–1844. In 1847 he was appointed by Congress to survey the mineral lands of Michigan. For biography, see the article by George P. Merrill and John F. Fulton in *DAB;* J. B. Woodworth in *The American Geologist,* August 1897, with an incomplete bibliography; or *CAB,* 3:97. For the controversy over the discovery of the use of ether in surgery, see Martin Gay, *Statement of the Claims of Charles T. Jackson to the Discovery of the Applicability of Sulpheric Ether to the Prevention of Pain in Surgical Operations* (Boston, 1847); J. L. Lord and H. C. Lord, *A Defense of Dr. Charles T. Jackson's Claims to the Discovery of Etherization* (Boston, 1848); or R. M. Hodges, a *Narrative of Events Connected with the Introduction of Sulpheric Ether into Surgical Use* (Boston, 1891). The Massachusetts Historical Society, Boston, has the Jackson papers.

WALTER R. JOHNSON (1794–1852) was born in Leominster, Massachusetts. He was principal of the High School of the Franklin Institute of Philadelphia, where he also taught chemistry and natural philosophy from 1826 to 1837. From 1839 to 1843 he was professor of chemistry and physics at the University of Pennsylvania. When the Association for the Advancement of Science was formed, he became its first secretary. Johnson's papers, 1831–1842, were mostly in physics-chemistry, with a few in natural history subjects and in meteorology. He was particularly interested in the problems of strength of materials and the relative efficiency of fuels. In 1844, under the authority of Congress, he made a number of investigations into the character of different coals. For biography, see Drake's *DAB;* some references in Joseph S. Hepburn, "Notes on the Early Teaching of Chemistry in the University of Pennsylvania, the Central High School of Philadelphia, and the Franklin Institute of Pennsylvania," *The Barnwell Bulletin,* X; No. 41 (October, 1932).

Isaac Lea (1792–1886) was born in Wilmington, Delaware of Quaker parents. He entered business with his brother in Philadelphia at the age of 15, and a few years later, in 1821, became a partner in the publishing house of his father-in-law, Matthew Carey. More than half of his publications, 1818–1839, were in zoology; the remaining were in other fields of natural history. His work was predominantly descriptive, but he was the first (1833) to transplant to America Lyell's terms, Pliocene, Miocene, and Iocene. His *Observations of the Genus Unio,* 1827–1874, in 13 volumes, described more than 1800 species of mollusca, recent and fossil. Lea was president of the Academy of Natural Sciences of Philadelphia (1858–1863), and vice-president of the American Philosophical Society. For a complete bibliography of works up to 1876, see his *Catalogue of the Published Works of Isaac Lea, LL.D. from 1817 to 1876* (Philadelphia, 1876). For biography, see J. H. and C. H. Lea, *The Ancestry and Posterity of John Lea* (Philadelphia and New York, 1906); Youmans, pp. 260–69; or N. P. Scudder's article in *Bulletin 23,* U.S. National Museum, 1885, which contains a complete bibliography and biographical sketch. His papers are in the Academy of Natural Sciences of Philadelphia.

John E. Leconte (1784–1860) was born in Shrewsbury, New Jersey, of a substantial family which included several other scientists. He was with the Corps of Topographical Engineers of the United States Army from 1818 to 1831, becoming a brevet-major in 1828. At one time he served as vice-president of the Academy of Natural Sciences of Philadelphia, and he was a member of the American Philosophical Society. His publications were in zoology and botany, almost entirely of a descriptive nature. For biography, see Drake's *DAB* or *The Autobiography of Joseph Leconte* (nephew), (New York), 1903, *passim,* for occasional mention. The Leconte papers are in the American Philosophical Society Library.

Charles Alexander Lesueur (1778–1849) was born in France, the son of an admiralty officer. His early training was in drawing and painting, and it was with these skills that he made his greatest contributions to natural history. He did pioneer work in natural history illustrations in France; and after coming to the United States as traveling companion of William Maclure, he became, next to Say, the most active zoologist of the time. His publications, 1817–1825, were all in zoology, with one exception, an article in paleontology. All his work was descriptive. During his residence in America, from 1815 to 1837, before returning to France to teach painting, he described two genera and fifteen species of American fishes, was the first to study the fishes of the Great Lakes, and prepared the illustrations for Godman's *American Natural History* and Say's conchology and entomology works. He was a member of the American Philosophical Society and the Academy of Natural Sciences of Philadelphia, and he was involved, with Maclure and others, in the New Harmony Movement. For biography, see David Starr Jordon's article in *DAB;* Youmans, pp. 128–39; or Adrien Loir, *Charles Alexander Lesueur Artiste et Savant Français en Amérique de 1816 à 1839* (Le Havre, 1920). For a complete bibliography, see *AJS*, VIII (1849), 189.

JOHN LOCKE (1792–1856) was born in Fryeburg, Maine, the son of a skilled mechanic and machine constructor. He attended the Academy of Bridgeport and, in 1816, began the study of medicine, receiving an M.D. from Yale. He served for a period as surgeon's mate in the navy. In 1818 he lectured on botany at Dartmouth, and the same year served as curator of the botanical gardens at Cambridge. He then taught at the Female Academy in Cincinnati until 1835, when he became professor of chemistry in the Ohio College of Medicine, a post he held for the remainder of his life, except for a brief interval when a dispute among the faculty forced him into temporary retirement. In 1838 he was employed as an assistant in the State Geological Survey. Locke published, 1820–43, articles in physics-chemistry, and a few each in several natural history subjects and one in astronomy. In 1819 he published *A Manual of Botany*, the illustrations for which he engraved himself. Locke was best known for his inventions of chemical and meteorological equipment, of an improved, portable botanic press, and most important of all, the electrochronograph or magnetic clock, in 1848. Locke was a member of the American Philosophical Society. For biography, see M. B. Wright, *An Address on the Life and Character of the Late Professor John Locke* (Cincinnati, 1857); Carl W. Mitman's article in *DAB;* and J. G. Locke, *Book of the Lockes* (Boston, 1853).

ELIAS LOOMIS (1811–1889) was born in Wilmington, Connecticut, the son of a country pastor. He was graduated from Yale in 1830, and from 1831 to 1832 attended the Andover Theological Seminary. He returned, however, to natural science, becoming a tutor at Yale under Denison Olmsted. It was at Yale, in 1835, that he and Olmsted observed Halley's Comet through the five-inch telescope, then the largest in the country. In 1836–1837 he studied at Paris under Arago, Biot, and Pouillet. From 1840 to 1844 he was professor of mathematics and natural philosophy at Ohio Western Reserve College, where he established an observatory; and from 1844 to 1860 he was professor of natural philosophy at the University of the City of New York, at which time he became professor of natural philosophy and astronomy at Yale, a position he held for the rest of his life. Loomis was a member of the American Philosophical Society and of the National Academy of Sciences. His publications, 1836–1844, were in astronomy, physics, and meteorology. He was the author of numerous mathematical textbooks, both introductory and secondary, as well as textbooks in astronomy and physics. For biography, see H. A. Newton's memoir in NAS *Biog. Mem.*, *III*, 1895, 215, with a complete bibliography; David E. Smith's article in *DAB;* Elias Loomis, *The Descendents of Joseph Loomis, Who Came from Braintree, England, in the Year 1638, and Settled in Windsor, Connecticut, in 1639*, 3rd edition, revised by E. S. Loomis (Berea, Ohio, 1909).

WILLIAM MACLURE (1763–1840) was born in Scotland, and after being educated by a well-known Scottish tutor engaged in the mercantile business. In 1796 he moved to the United States, with his fortune already made, where he became both a patron of science and an active scientist himself. Maclure was a founding member of the Academy of Natural Sciences of Philadelphia and served as President of the Academy from 1817 to 1840. During his affiliation with the Academy, he donated over $20,000 to provide a permanent building. Among other activities, he persuaded Joseph Neef to come to America to introduce Pestalozzian educational methods, brought Lesueur to the United States, and purchased the land for the New Harmony movement, after the failure of a similar try in Spain, which he also financed. Maclure was a member of the American Philosophical Society. His publications, 1817–1829, were in geology and educational theory. His best-known work was *Observations on the Geology of the United States of North America* (1817). For biography, see George P. Merrill's article in *DAB;* Samuel George Morton's in *AJS*, XLVII (1844), 1–17; or Charles Keyes, "William Maclure, Father of American Geology," *Pan-American Geologist*, XLIII (1925), 81–94.

SAMUEL LATHAM MITCHILL (1764–1831) was born in Long Island, New York, and in 1786 received an M.D. from Edinburgh, after having studied with Samuel Bond in New York City. From 1792 to 1800 he was professor of chemistry and natural philosophy at Columbia College; from 1801 to 1826 professor of botany and materia medica at the same institution. His publications, between 1818 and 1828, covered the entire range of natural history, but unlike most men engaged in natural history at the time, he was marked by a speculative boldness and originality that led him, for example, to endorse and write a commendatory preface to the American edition of Erasmus Darwin's *Zoonomia* (1793). In addition to his writing, Mitchill was important in the founding of a number of scientific institutions of the period; including the first strictly scientific journal in America, the *Medical Repository* (1797), which he edited until 1813; the Society for the Promotion of Agriculture, Arts and Manufactures, under whose auspices he made a pioneering mineralogical expedition to the banks of the Hudson River in 1796; and the Rutgers Medical School (1826). Mitchill was a member of the American Philosophical Society. The best source for his biography and account of his works is Courtney R. Hall, *A Scientist in the Early Republic, Samuel Latham Mitchill, 1764–1831* (New York, 1934). For further references, see Bell.

SAMUEL GEORGE MORTON, 1799–1851, was born in Philadelphia and studied at Friends boarding schools and at the private school of John Gummere in Burlington, New Jersey. He studied in Philadelphia under Doctor Joseph Parrish and received an M.D. from the University of Pennsylvania in 1820, and from Edinburgh in 1823. He also studied medicine for a short period in Paris. Although he taught briefly at a number of medical schools, he spent most of his adult life actively engaged in research. His publications between 1829 and 1844 were mostly in paleontology, anthropology, and geology, but his major research interest was in collecting a large series of human skulls for comparison. According to the catalogue of 1849, his collection of skulls contained 1,512 specimens, of which about 900 were human. This was the largest collection in the world at that time. Largely on the basis of his studies of skulls, when the controversy over the origin of the races began he defended the multiple-origin theory. He was a member of the American Philosophical Society, the American Academy of Arts and Sciences, and served a term as president of the Academy of Natural Science of Philadelphia. Morton's papers are in the American Philosophical Society and in the Library Company of Philadelphia. For biography, see William R. Grant, *Sketch of the Life and Character of Samuel George Morton, M.D., Lecture Introductory to A Course on Anatomy and Physiology in the Medical Department of Pennsylvania College, Delivered, October 13, 1851* (Philadelphia, 1852); Gross' *Lives; DAB;* George B. Wood, *A Biographical Memoir of Samuel George Morton, M.D.* (Philadelphia, 1853); Charles D. Meigs, *A Memoir of Samuel George Morton, M.D.* (Philadelphia, 1851), containing a full list of his writings; Henry S. Patterson, "Memoir of the Life and Scientific Labors of Samuel George Morton," in Josiah C. Nott and George R. Gliddon, *Types of Mankind; Or Ethnological Researches* (Philadelphia, 1854), pp. xvii–lvii.

THOMAS NUTTALL (1786–1859) was born in Yorkshire, England, and was a printer by profession before turning to natural history. He was not formally educated, but was instructed in botany by Benjamin Smith Barton after his arrival in America. From 1822 to 1834 he was curator of the Harvard Botanical Garden, and in 1834 he returned to England to live for the remainder of his life, having inherited an estate on the condition that he reside in England. During his stay in America he toured the entire eastern and midwestern portions of the country and traveled to California collecting natural history specimens, usually on foot or by canoe. His publications, from 1817 to 1837, were largely in botany, with a scattering in other natural history subjects. His approach was almost purely descriptive, but in one article (ANS Philadelphia, *Journal*, 1821, II, 14 ff.) he did attempt to correlate geological formations widely separated geographically by means of the fossils found in them (secondary formation of Iowa and the mountain limestone of Derbyshire). Nuttall was a member of the American Philosophical Society, the Academy of Natural Science of Philadelphia, and the Linnaean Society of London. The Academy of Natural Science of Philadelphia has some of Nuttall's letters in its miscellaneous manuscripts collection, and the *Popular Science Monthly* for March, 1895, has a complete Nuttall bibliography. There is an autobiographical sketch in the Preface to vol. IV (1842) of his *North American Sylva*. For biography, see Youmans, pp. 205–14; J. W. Harshbarger, *The Botanists of Philadelphia* (Philadelphia, 1899); William E. Leonard, "Some Early Philadelphia Botanists—Schweinitz, Nuttall, Rafinesque, and Darlington," Minnesota Academy of Science, *Bulletin*, XIII (1887), 29–37; Richard G. Beidleman, "Some Biographical Sidelights on Thomas Nuttall, 1786–1859," APS *Proc.*, 104 (1960), 86–100; and Winmer Stone, "Some Philadelphia Ornithological Collections and Collectors, 1784–1850," *The Auk*, n.s., XVI (1899), 166–67. The sketch by C. R. Keyes in the *Popular Science Monthly*, for March, 1895, contains a complete bibliography. For further references, see Bell.

DENISON OLMSTED (1791–1859) was born in East Hartford, Connecticut, the son of a farmer. He was graduated from Yale in 1813, and from 1815 to 1817 he was a tutor and student of theology at Yale. From 1817 to 1824 he was professor of chemistry, mineralogy, and geology at the University of North Carolina; from 1825 to 1836, professor of mathematics and natural philosophy at Yale. In 1836 at his request the chair at Yale was divided and he held the professorship of natural philosophy for the rest of his life. In 1821 he conducted a geological survey of North Carolina—the first to be sponsored by a state legislature. Olmsted first gained scientific fame with papers dealing with the meteoric showers of November 13, 1833, in *AJS* from 1834 to 1836, in which he suggested an annual occurrence. His publications, 1820–1839, all in *AJS*, were predominantly in chemistry and meteorology, but he also did work in geology, astronomy, and mineralogy. For biography, see Youmans, 259–69; the article by Alois F. Kovarik in *DAB*; Franklin B. Dexter, *Biographical Sketches of the Graduates of Yale College*, IV (New York, 1912), containing a full bibliography; the article by T. D. Woolsey in the *New Englander* for August, 1849; C. S. Lyman's article in *AJS* for July,

1859; K. P. Battle, *History of the University of North Carolina* (2 Vols., 1907–1912); and H. K. Olmsted and G. K. Ward, *Genealogy of the Olmsted Family in America* (New York, 1912).

CHARLES G. PAGE (1812–1868) was born in Salem, Massachusetts. Graduated from Harvard in 1832, he became a medical student in Boston, and in 1838 was practicing medicine in Virginia. In 1840 he became professor of chemistry at Columbia University. All of his publications, between 1834 and 1839, were in chemistry, and almost all were on electromagnetism as a motive force, his special interest. From 1841 to 1868 Page worked as an examiner in the United States Patent Office. He made a great number of improvements on the electrical machines of the day, including an induction apparatus of greater intensity than Henry's self-acting circuit breaker. In 1851 he built a locomotive using two of his electric engines, which, with all of his improvements, were probably the best of that time. When he tried it out on a five-mile track from Washington to Bladensburg, Maryland, it attained a speed of 19 miles per hour, but the electric batteries could not furnish the current required to operate the locomotive for any appreciable length of time. For his own account of his electromagnetic work, see his *History of Induction: The American Claim to the Induction Coil and Its Electro-Static Development* (Washington, 1867), also published in the *American Polytechnic Journal,* which he edited from 1853 to 1854. His *Psycho-Mancy, or Spirit Rappings Exposed* (New York, 1853) was a major American attack on the great wave of spiritualism that swept the country in the 1850s. Page was a member of the American Academy of Arts and Sciences. For biography, see P. S. W. Page, *Reminiscences, 1883–1886* (New York, 1886); C. N. Page, *Genealogical Chart of the Page Family* (Des Moines, 1911); Drake's *DAB;* and the article by Carl W. Mitman in *DAB.*

CONSTANTINE S. RAFINESQUE (1773–1840) was born in Constantinople, the son of a French merchant. He never held down a job for long, but preferred the life of a flamboyant traveler and collector. His publications, all in 1818, were in descriptive zoology, botany, and meteorology. One of the most active naturalists of his time, he has been credited with establishing thirty-four genera and twenty-one species of American fishes. His work, *The Medical Flora of North America* (Philadelphia, 1828) was an important influence on the botanic medicine movement. Because of his picturesque character, he has been extensively studied. A statement he made in his *Atlantic Journal* (Spring, 1833), p. 164, has led some to consider him an early evolutionist. For biography, see Richard E. Call, *The Life and Writings of Rafinesque* (Louisville, 1895), or, by the same writer, an edition of Rafinesque's *Ichthyologia Ohiensis with a Sketch of Rafinesque* (Cleveland, 1899); T. J. Fitzpatrick, *Rafinesque: A Sketch of His Life* (Des Moines, 1911); E. D. Merrill, in "A Generally Overlooked

Rafinesque Paper," APS *Proc.*, LXXXVI (1943), 72–90, adds 34 titles to Fitzpatrick's list of Rafinesque's works and he adds to Fitzpatrick's list of works on Rafinesque; Harry B. Weiss, *Rafinesque's Kentucky Friends* (Highland Park, New Jersey, 1936) includes some of Rafinesque's unpublished drawings. For further reference, see Bell.

WILLIAM C. REDFIELD (1789–1852) was born in Middleton, Connecticut, where he was apprenticed to a saddle and harness maker at an early age. He was self-taught in science and his profession was that of a naval engineer. Redfield was a member of the American Philosophical Society, the American Academy of Arts and Sciences, and when the American Association for the Advancement of Science was formed in 1848, he became its first president. His publications, 1831–1843, in AJS, were mostly in meteorology, but he also did some work in descriptive paleontology. He was primarily known for his close studies of hurricanes and other storms, which led to his discovery of the rotary motion of winds in storms. After the period covered by this study, he devoted all his scientific work to paleontology, becoming the first American specialist on fossil fish. For biography, see the article by William J. Humphreys in *DAB;* the sketch by Denison Olmsted in *AJS*, XXIV (1857), 355–73, containing a bibliography; or J. H. Redfield, *Genealogical History of the Redfield Family* (New York, 1860). For his work in paleontology, see George G. Simpson, "The Beginnings of Vertebrate Paleontology in North America," *Proc.* APS, LXXXVI (1943), 167. Yale University has three volumes of his letterbooks.

JEREMIAH VAN RENSSELAER (1793–1871) was born in Greenbush, New York, the son of General J. J. Van Rensselaer. He studied under his uncle, Archibald Bruce, for a while in New York, received an M.D. from Edinburgh after graduating from Yale in 1813, and practiced medicine in New York City. His publications, 1820–1842, covered the range of natural history. For biography, see *CAB*, 7:525.

THOMAS SAY (1787–1834), a druggist by profession, was born in Philadelphia, the son of a prominent physician. His works, 1817–1839, were almost entirely in descriptive zoology and paleontology, but he was the first American to point out the chronogenetic value of fossils (*AJS*, I, 1818, p. 381). Say was the zoologist for Long's expedition to the Rocky Mountains in 1819 and to the Minnesota River in 1823. From 1822 to 1828 he was professor of natural history at the University of Pennsylvania. He was involved with Maclure and others in the New Harmony movement. Say was a member of the American Philosophical Society. For biography, see H. B. Weiss and Grace M. Zigler, *Thomas Say, Early American Naturalist* (Baltimore: C. C. Thomas, 1931); J. L. Leconte's edition of *The Complete Writings of Thomas Say on the Entomology of North America* (New York and London, 1859), which contains a biographical memoir by George Ord; Leland Howard's article in DAB; or Harry B. Weiss, *The Pioneer Century of American Entomology* (New Brunswick, New Jersey, 1936). G. D. Harris reprinted his paleontological writings in *Bulletin of American Paleontology*, I, No. 5, 1896. W. G. Binney has edited *The Complete Writings of Thomas Say on the Conchology of the United States*. For further reference, see Bell.

HENRY ROWE SCHOOLCRAFT, 1793–1864, was born in Watervliet, New York, the son of a glassmaker. He attended Middlebury College and Union College. From 1822 to 1864 he served as Indian agent and in other capacities in government service on the Northwestern frontier. In 1846 he received an LL.D. from Geneva. His periodical publications, from 1815 to 1824, were in geology, paleontology, and mineralogy, in addition to one article on glass manufacturing. His chief work was *History and Statistical Information Respecting the History, Conditions and Prospects of the Indian Tribes of the United States*, 6 vols. (Washington, 1851–1857), which was paid for by Congress. Schoolcraft was a member of the American Philosophical Society. For biography, see C. S. and S. Osborn, *Schoolcraft, Longfellow and Hiawatha* (Lancaster, Pennsylvania, 1942), with a Schoolcraft bibliography; Youmans, pp. 300–310; and the article by Walter Hough in *DAB*.

HENRY SEYBERT (1801–1883) was born in Philadelphia, the son of Adam Seybert, a noted scientist. His early education was conducted by his father, whose traveling companion and scientific assistant he became. He studied at the Ecole des Mines, Paris, and upon his return to Philadelphia at the age of twenty-one was made a member of the American Philosophical Society. Between 1821 and 1825 he published 15 chemical analyses of minerals. Although he lived to old age, he was not very productive after the death of his father in 1825. The last half of his life he spent as a philanthropic promoter of human welfare and of interest in science. He became interested in psychical research in the 1830s, and on his death bequeathed to the University of Pennsylvania the sum of $60,000 to endow a chair of philosophy on the condition that an additional sum be used to support the activities of a commission of the University appointed to investigate modern spiritualism. For biography, see Moncure Robinson in APS *Proc.*, XXI (1884), 241–63; E. P. Oberholtzer, *Philadelphia: A History of the City and Its People* (Philadelphia, 1912), vol. 2; an obituary notice in the *AJS*, XXV (1883); and the article by Courtney R. Hall in *DAB*.

CHARLES U. SHEPARD (1804–1886) was born in Newport County, Rhode Island, the son of a minister. He studied at Brown University in 1820–1821, and in 1824 graduated from Amherst. He then studied under Thomas Nuttall for a year before becoming assistant to Silliman at Yale. By the time of his death, his collection of minerals was reported to have been the largest in America. Shepard was lecturer on natural history

at Yale from 1830 to 1847; professor of chemistry and natural history at
Amherst, 1835 to 1852; and professor of chemistry at the Medical
College of Charleston, South Carolina, from 1834 to 1861. In 1835 he
was an associate of Percival in the State Geological Survey of Connecticut. Shepard was a member of the American Academy of Arts and
Sciences. His publications, 1824–1843, were in mineralogy, chemistry,
and geology. His major work was *A Treatise on Mineralogy* (1835). For
biography, see Youmans, pp. 419–27; the article by Frederick B. Loomis
in *DAB;* and the *New England Historical and Genealogical Register* for
April, 1869.

BENJAMIN SILLIMAN (1779–1864) was born in North Stratford (now
Trumbull), Connecticut, of a prominent family, his father having been a
Revolutionary general. He graduated from Yale in 1796, then studied law
and tutored at Yale. In 1802 he was appointed professor of chemistry and
natural history at Yale, with permission to qualify himself before assuming the duties of the position. He studied for two years with James
Woodhouse, two years at Edinburgh, and in 1808 assumed his profes-
sorial duties at Yale. Silliman received an M.D. from Bowdoin in 1818
and an LL. D. from Middlebury College in 1826. In 1818 he founded the
American Journal of Science and Arts, which he edited until his death.
His publications, 1820–1842, were mostly in chemistry and geology. Silliman was greatly concerned with the problem of scientific education and
in the problem of reconciling science with the Bible, in which he devoutly
believed, although he was not a fundamentalist. Silliman was a member
of the American Philosophical Society, the American Academy of Arts
and Sciences, and a number of foreign scientific societies. For biography,
see John F. Fulton and E. H. Thomason, *Benjamin Silliman, 1779–1864,
Pathfinder in American Science* (New York, 1947). Yale University has
his papers, diary, and letterbooks, along with nine volumes in manuscript
on "The Origin and Progress of Chemistry, Mineralogy and Geology in
Yale College"; the Historical Society of Pennsylvania has about 500 of his
letters; and the Edgar Fahs Smith Memorial Collection of the University
of Pennsylvania has a large collection of his correspondence with Robert
Hare.

J. LAWRENCE SMITH (1818–1883) was born in Charleston, South Carolina, the son of a merchant. After graduating with a classical education
from Charleston College, he studied science at the University of Virginia.
Smith worked as a civil engineer for two years, then studied medicine at
the Medical College of Charleston. From 1841 to 1844 he was abroad,
studying physiology under Flourens and Longet; chemistry under Orfia,
Dumas, and Liebig; physics under Pouillet, Desprez, and Becquer; and
mineralogy under Elie de Beaumont and Dufrenoy. He served briefly as a
professor at a number of colleges, including a term as professor of
chemistry at the University of Virginia (1850), and four years at the

University of Louisville (1852–1856), but spent most of his life as the president of a chemical company. In 1846, with S. D. Sinkler, he established the *Southern Journal of Medicine and Pharmacy*. From 1846 to 1850 he was a mining engineer for the Turkish government. His publications, 1839–1844, were all in chemistry and his particular interest was in organic chemistry. Although at the time of his death he had one of the world's largest private collections of meteorites, he did no significant work with them. Smith was a member of the American Philosophical Society, the American Medical Association, and served one term as president of the American Academy of Arts and Sciences. For biography, see *CAMB;* the article by T. Carey Johnson, Jr. in *DAB;* J. B. Marvin (editor) *Original Researches in Mineralogy and Chemistry by Professor J. Lawrence Smith* (Louisville, 1884), which has a list of his published papers and biographical sketches by Marvin, Silliman, Jr., and Middleton Michel; or the obituary in *AJS*, XXVI (1883), 414–15.

D. Humphreys Storer (1804–1891) was born in Portland, Maine, the son of a shipowner, merchant, and collector of customs at Portland. He graduated from Bowdoin College in 1822, then began the study of medicine with John C. Warren and received an M.D. from the Harvard Medical School in 1825. From 1825 to 1888 he practiced medicine in Boston. Becoming dissatisfied with the four months' winter term offered at the Harvard Medical School, in 1839, in association with O. W. Holmes, Edward Reynolds, and Jacob Bigelow, he helped found the Tremont Street Medical School, holding classes throughout the year. This institution flourished until the Harvard School was reorganized. From 1854 to 1868 Storer was professor of obstetrics and medical jurisprudence at Harvard. His publications, from 1836 to 1842, were all in descriptive zoology, especially on fishes and reptiles; and most were published in the *Boston Journal of Natural History*, organ of the Boston Society of Natural History, an organization which he helped found. In 1837 Storer was appointed zoologist and herpetologist for the Massachusetts geological survey. Storer was a member of the American Philosophical Society and the American Academy of Arts and Sciences. For biography, see William H. Shastid's article in *CAMB;* the article by Henry R. Viets in DAB; S. H. Scudder in *Proc. AAA&S*, XXVII (1891), 388–91; Malcolm Storer, *Annals of the Storer Family* (Boston, 1927); T. F. Harrington, *The Harvard Medical School*, II (New York and Chicago, 1905); or G. C. Shattuck, O. W. Holmes, and others, *Boston Medical and Surgical Journal*, March 24, 1892. A "List of the Published Writings of David Humphreys Storer" is in Bowdoin College Library, *Bibliographical Contributions*, No. 2, August, 1892. A volume of his bound manuscript letters is in the Museum of Science, Boston.

Theodore Strong (1790–1869) was born in South Hadley, Massachusetts, the son of a Congregational minister. He received his early education from his uncle, Colonel Woodbridge, and graduated from Yale in 1812. He was professor of mathematics and natural philosophy at Hamilton College from 1827 to 1861. All of his published works, 1818–1843, were in mathematics, and he was particularly interested in problems of the composition and resolution of forces. He made no original contributions to this field, however, but was better known as an effective and inspiring teacher and writer of textbooks. Strong was a member of the American Philosophical Society and the American Academy of Arts and Sciences. For his work as a teacher, see Florian Cajori, *The Teaching and History of Mathematics in America* (Washington, 1890), *passim.* For biography, see the memoir by Joseph P. Bradley in NAS *Biog. Mem.,* I (1886), 1; the article by Raymond C. Archibald in *DAB* or B. W. Dwight, *The History of the Descendants of the Elder John Strong* (Albany, New York, 1871), 2 vols. AAA&S *Proc.,* VIII (1873) contains a bibliography of Strong's work.

JOHN TORREY (1796–1873) was born in New York City, of an old substantial family. He received an M.D. from the College of Physicians and Surgeons of New York University, and began practicing medicine in New York in 1820. In 1824 he was appointed assistant surgeon in the United States Army and professor of chemistry at West Point. From 1827 to 1854 he was professor of chemistry at the College of Physicians and Surgeons, New York, and from 1830 to 1854 also served as professor of chemistry and natural history at the College of New Jersey; from 1854 to 1873 he served as United States Assayer. His journal publications, 1820–1837, were predominately in botany, with some works in chemistry and mineralogy. Torrey was president of the American Academy of Arts and Sciences and a corporate member of the National Academy of Science. For biography, see Andrew D. Rodgers, *John Torrey, A Story of North American Botany* (Princeton, 1942); H. A. Kelly, *Some American Medical Botanists* (New York, 1929), pp. 136–44; and the memoir by Asa Gray in NAS *Biog. Mem.,* I, 267.

GERARD TROOST (1776–1850) was born in Holland and received an M.D. from the University of Leyden, and a Master of Pharmacy degree from the University of Amsterdam in 1801. Between 1807 and 1810 he studied under Haüy and Werner; then he came to Philadelphia, where in 1817 he became one of the founders of the Academy of Natural Sciences. In 1821 he was appointed professor of mineralogy in the Philadelphia Museum, holding this position until 1825, when he joined the group moving to New Harmony. From 1828 to 1830 he served as professor of chemistry, geology, and mineralogy at the University of Nashville, and from 1831 to 1850 he also served as geologist and mineralogist to the state of Tennessee. Troost was a member of the American Philosophical Society. His publications, 1825–1840, were almost entirely in the field of descriptive mineralogy, displaying the interest in crystalline forms he learned from Haüy; but he also published two chemical analyses of minerals and one work in zoology. The Troost collection of minerals, numbering over 13,000 specimens, is now in the Museum of the Louisville Free Library. For a consideration of Troost's methods in natural history, see Elvira Wood, *A Critical Summary of Troost's Unpublished Manuscript on the Crinoids of Tennessee* (Washington, 1909). For biography, see Youmans, pp. 119–28; L. C. Glen in the *American Geologist* for February, 1905, which includes (pp. 90–94) a list of Troost's published writings; Philip Lindsley, "The Life and Character of Professor Gerard Troost, M.D.," in *The Works of Philip Lindsley*, vol. I (Philadelphia, 1866); and H. G. Rooker, "A Sketch of the Life and Work of Dr. Gerard Troost," *Tennessee Historical Magazine*, October, 1932. The diary of J. B. Lindsley now in the Peabody College Library contains many references to Troost.

LARDNER VANUXEM (1792–1848) was born in Philadelphia, the son of a prosperous shipping merchant. He was graduated from the Paris School of Mines in 1819 and from 1819 to 1826 served as professor of chemistry and mineralogy at South Carolina College. He then worked as a mining engineer in Mexico, and, in 1830, purchased a farm near Bristol, Pennsylvania, where he lived the rest of his life in seclusion, except for the four-year period 1836–1840 during which he was employed in the New York Survey. It is the work that came out of this survey, *Geology of New York, Third District* (New York, 1842), on which his scientific reputation rests. In this work, he brought forward views which seem impossible to harmonize with the mosaic account of creation, and he appears faintly to have postulated some kind of theory of evolution (p. 27). It was his suggestion, in 1838, that geologists of New York, Pennsylvania, and Virginia meet to adopt a nomenclature for American geology that led to the formation of the Association of American Geologists. His publications, 1821–1829, were in chemistry, mineralogy, and geology. Vanuxem was a member of the American Philosophical Society. According to Youmans (p. 275), he left an "immense pile" of manuscript books on his religious studies, but I have not been able to locate these. For biography, see Youmans, pp. 270–78; the article by F. Lynwood Garrison in *DAB;* the obituary notice in *AJS,* V (1848), 445–46; and Maximilian La Borde, *History of the South Carolina College* (Charleston, 1874).

JOHN WHITE WEBSTER (1793–1850) is primarily famous as the unsuccessful defendant in one of the most notorious murder trials of the nineteenth century. In 1850 he was hanged for murdering his Harvard colleague, George Parkman, and disposing of all of his body—except his Harvard-made teeth, which survived as evidence. Webster was born in Boston, the son of an apothecary, and he received an M.A. from Harvard in 1811 and an M.D. from the same school in 1815. Afterwards, he studied at Guy's Hospital, London, and from 1824 to 1829 he served as professor of chemistry at Harvard. From 1823 to 1824 he was co-editor of the *Boston Journal of Philosophy and Arts*. His publications, 1818–1839, were in mineralogy, geology, chemistry, paleontology, and natural history. Webster's conviction in the murder trial touched off a long pamphlet controversy; a collection of these pamphlets may be found in the Library of Congress, the Boston and New York Public Libraries, the Boston Athenaeum, the American Antiquarian Society, and the Harvard Library. For biography, see Edmund L. Pearson in *DAB;* H. B. Irving, *A Book of Remarkable Criminals* (New York, 1918); Drake's *DAB;* "Harvard Homicide," *MD M. Mag.,* 4 (3), 1960, pp. 182–93. Because of the spectacular nature of Webster's crime, none of the biographical sources have had anything to say about his scientific work. He was, however, a first-rate investigator and reporter in geology and mineralogy. See particularly, *A Description of the Island of St. Michael* . . . (Boston, 1821). He was also an early contributor to the *North American Review.*

JEFFRIES WYMAN (1814–1874) was born in Chelmsford, Massachusetts, the son of the physician to the McLean Asylum for the Insane at Somerville, Massachusetts. He graduated from Harvard in 1833 and received his M.D. from the Medical College of Boston in 1837, after which time, he became demonstrator in anatomy under John C. Warren. He was curator of the Lowell Institute in 1840–1841, delivering a course of lectures there in addition to his other duties, and in 1841 he studied in Paris. In 1843 he became professor of anatomy and physiology in the medical department of Hampden Sidney College, Hampden Sidney, Virginia, and in 1847 succeeded Warren as Hersey Professor of Anatomy at Harvard. While at Harvard he built up the museum of comparative anatomy. In 1847 he became the first to give a scientific description of the gorilla, and his special studies of temporary organs and useless organs in animals were cited by A. S. Packard as contributions to evolution theory, which he apparently did embrace, although he did not enter into the controversy. During the period of this study he was primarily an anatomist, but later in life he turned to anthropology and became the leading anthropologist in America by the time of his death. His publications, 1839–1844, were in physiology, paleontology, and zoology. Wyman was a member of the American Academy of Arts and Science, serving as its president for one term; the Association of American Geologists and Naturalists; The American Philosophical Society; and was appointed a corporate member of the National Academy of Science, but resigned the position for an unexplained reason. For biography, see A. S. Packard's memoir in NAS *Biog. Mem.,* II (1886), 77, which includes a complete bibliography; Asa Gray's article in *Proceedings of the Boston Society of Natural History,* XCII (1874), 95–125; O. W. Holmes in the *Atlantic Monthly* for November, 1874, pp. 611–23; B. G. Wilder's sketch in *Popular Science Monthly,* VI (1875), 355–60; F. W. Putnam in *Proc. AAA&S,* XI (1876), 496–505, with a bibliography; S. Weir Mitchell in *Lippincott's Magazine* for March 1875; Hubert L. Clark in *DAB;* T. F.

Harrington, *The Harvard Medical School* (New York, 1905) 815–31, and *A History of the First Half Century of the National Academy of Sciences* (Washington, 1913), 197–200. For a recent estimate of his views on evolution, see A. Hunter Dupree, "Jeffries Wyman's Views on Evolution," *Isis*, 44 (1953), 243–46, and by the same author, "Some Letters from Charles Darwin to Jeffries Wyman," *Isis*, 42 (1951), 104–10. See also, Raymond N. Doetsch, "Early American Experiments on 'Spontaneous Generation' by Jeffries Wyman (1864–1874)," *Journal of the History of Medicine and Allied Sciences*, XVII (July, 1962), 325–32.

APPENDIX II

❦

American Scientific Journals,

1771–1849

The following tables were compiled from a list of periodical publications collected during the course of this research. They include all those publications (337 titles) which, to my knowledge, made a regular practice of including scientific material, either original research or abstracts of scientific work published elsewhere.

Table VII

Number and Distribution by States of Scientific Journals Founded, by Five-Year Periods, 1805–1849

States	Prior to 1805	1805 to 1809	1810 to 1814	1815 to 1819	1820 to 1824	1825 to 1829	1830 to 1834	1835 to 1839	1840 to 1844	1845 to 1849
Penn.	6	4	4	3	5	10	6	11	7	7
N.Y.	4	8	6	9	7	3	9	8	19	20
Mass.	4	2	3	–	5	5	6	11	11	9
N.J.	2	–	–	1	–	1	–	–	–	2
Md.	1	2	1	1	2	1	4	1	2	–
S.C.	1	–	–	–	–	1	–	1	–	2
Conn.	1	–	2	1	2	–	–	1	3	–
Ga.	1	–	–	–	–	–	–	2	2	4
D.C.	–	–	1	–	1	1	1	–	2	2
Vt.	–	–	1	–	–	–	1	2	1	1
Ky.	–	–	–	1	1	2	–	3	2	2
Ohio	–	–	–	–	1	4	6	6	4	8
Tenn.							1	1	–	1
La.							1	1	1	2

Table VII (Continued)

States	Prior to 1805	1805 to 1809	1810 to 1814	1815 to 1819	1820 to 1824	1825 to 1829	1830 to 1834	1835 to 1839	1840 to 1844	1845 to 1849
Maine								2	–	–
Mich.								1	–	2
Miss.								1	–	1
R.I.								2	1	–
Ill.									1	1
Mo.									1	2
Va.									1	–
Ind.										2
Wis.										1

Table VIII

Length of Time in Publication of Newly Founded Scientific Journals, by Five-Year Periods, 1805–1849

	Less Than 1 Year	1–2 Years	2–5 Years	5 or More Years	Total
Prior to 1805	5	3	1	11	20
1805–1809	3	3	3	6	15
1810–1814	8	2	4	4	18
1815–1819	5	5	1	6	17
1820–1824	8	1	7	7	23
1825–1829	8	5	4	9	26
1830–1834	20	8	4	4	36
1835–1839	18	5	14	15	52
1840–1844	22	11	10	13	56
1845–1849	25	8	15	6	74

Table IX

Scientific Journal Publication Data, 1805–1849

Year	Journals in Publication Total Number	Trans-actions of Medical Societies	Sectarian Medical Journals	Journals Founded	Journals Deceased
1805	12	4		1	0
1806	18	8		6	0
1807	21	11		3	3
1808	20	10		3	2
1809	19	10		1	2
1810	24	11		8	3
1811	23	11		2	3
1812	23	12		4	1
1813	24	12		2	3
1814	26	12		3	5
1815	21	11		2	1
1816	20	11		0	2
1817	23	12		5	3
1818	26	10		6	3
1819	27	10		3	3
1820	27	11		5	4
1821	24	11		1	1
1822	29	11	1	6	3
1823	31	11	2	5	4
1824	34	11	2	7	6
1825	34	11	0	6	5
1826	33	11	1	3	6
1827	36	11	3	9	6
1828	36	11	3	6	5
1829	34	10	3	4	3
1830	27	12	5	7	3
1831	37	12	4	4	7
1832	44	13	7	14	11
1833	37	12	5	6	4
1834	40	12	5	6	3

Table IX (Continued)

Year	Journals in Publication Total Number	Trans- actions of Medical Societies	Sectarian Medical Journals	Journals Founded	Journals Deceased
1835	50	11	16	15	11
1836	51	11	17	12	11
1837	48	11	15	7	3
1838	55	12	20	10	7
1839	57	12	17	7	9
1840	54	12	18	7	7
1841	61	14	20	11	4
1842	66	14	24	8	11
1843	69	14	21	12	16
1844	68	14	20	10	4
1845	76	14	20	12	11
1846	77	15	19	12	8
1847	90	17	20	19	12
1848	90	20	17	13	15
1849	94	21	18	18	9

Notes

Introduction

The following abbreviations are used for works frequently cited:

AAAS Proc.,   *Proceedings of the American Association for the Advancement of Science*
AJS   *American Journal of Science and Arts*
APS, Trans.,   *Transactions of the American Philosophical Society*
NA Rev.,   *North American Review*

1. P. W. Bridgmen, *Reflections of a Physicist*, 2d ed., enlarged (New York, 1955), p. 341.

2. The outstanding example of the doctrinaire progressive as historian of science is George Sarton. For a succinct statement of his views, see "The Quest for Truth: A Brief Account of Scientific Progress During the Renaissance," in Dorothy Stinson (ed.), *Sarton on the History of Science* (Cambridge, 1962), p. 102. The chief challenger to this viewpoint is Lynn Thorndike; see especially his *A History of Magic and Experimental Science* (New York, 1923–1958), eight volumes of which have been published. Aside from this general work, a few studies of specific sciences have been made outside an obvious developmental framework. See, for example, Charles C. Gillispie, *Genesis and Geology, A Study in the Relation of Scientific Thought, Natural Theology and Social Opinion in Great Britain, 1790–1850* (Cambridge, 1951), or William M. Stanton, *The Leopard's Spots: Scientific Attitudes toward Race in America, 1815–1859* (Chicago, 1960). Richard H. Shryock's *Medicine and Society in America, 1660–1860* (New York, 1960), is a rare combination of both aspects of science.

3. M. Pattison Muir, *A History of Chemical Theories and Laws* (London, 1909), p. viii.

4. For this argument, see Thomas S. Kuhn, *The Structure of Scientific Revolutions* (Chicago, 1962). For a briefer statement by Kuhn and comments by several historians of science, see A. B. Crombie (ed.),

Scientific Change (New York, 1963), pp. 347–98. Although my own research was completed before I read Kuhn's work, I am happy to acknowledge my indebtedness to him in the final revision of this study.

5. For an example of the distance between the innovative genius responsible for significant scientific change and his fellows, one could note that the eighth edition of the *Encyclopaedia Britannica* gave only passing and slighting notice to the mechanical theory of heat and persisted in the caloric notion. On this point, see Arthur Hughes, "Science in English Encyclopedias, 1704–1875," *Annals of Science*, VII (1951), 364–65. The main conclusion that Hughes draws from his detailed study appearing in three parts in the *Annals of Science* (VII, 340; VIII, 323; IX, 233) is that general acceptance, even among highly literate circles, of scientific changes is incredibly slow.

6. The distinction between points of view made here is essentially the same as that between the "epistemological" and the "anthropological" point of view, which was first pointed out by the Polish sociologists, Maria and Stanislav Ossowski in 1936 (in Polish); an English translation of their paper, entitled "The Science of Science," is reprinted in Norman Kaplan (ed.), *Science and Society* (Chicago, 1965), pp. 19–29.

7. In calculating the amount of publication of the scientists, I have generally accepted the definitions of the day concerning what is a "scientific journal" and what is a "scientific article." The first is a journal which the editors denominate scientific, and the second is an article found in this type of journal. However, I have excluded articles having to do with the art of medicine, as distinguished from any of the sciences of which it is composed, pharmaceutical recipe-type articles, and "how to" articles of a popular nature. This is justified on the grounds that my main concern is with the research interests of the scientists. A study by Wayne Dennis, "Bibliographies of Eminent Scientists," *Scientific Monthly*, LXXIX (1954), 180, indicates that the "greater the number of a scientist's publications, the higher his scientific reputation is likely to be." As Dennis concedes, there are certainly anomalies, but there is nevertheless a high correlation between quantity and eminence.

8. The journals used were: *American Journal of Science and Arts* (1818–1844); *American Medical Recorder* (1818–1829); *American Medical Review and Journal* (1824–1825); *American Journal of the Medical Sciences* (1827–1829); *American Journal of Pharmacy* (1829–1844); *Boston Journal of Natural History* (1830–1844); *Literary and Scientific Repository* (1820–1822); *The Medical Repository* (1815–1824); *Memoirs of the American Academy of Arts and Sciences* (1818, 1820, 1833); *Journal of the Academy of Natural Sciences of Philadelphia* (1817–1844); *New York Medical and Physical Journal* (1822–1829); *North American Medical and Surgical Journal* (1826–1831); *Philadelphia Journal of the Medical and Physical Sciences* (1820–1827); *Transactions of the American Antiquarian Society* (1820, 1837); *Transactions of the American Philosophical Society* (1815–1844); *West-*

ern Journal of the Medical and Physical Sciences (1827–1830). I have not, however, limited myself to those journals after making the initial selection.

I. The Pursuit of Science in America

1. "Presidential Address," *Proceedings of the Royal Society of London,* XXXIX (1885), 278.
2. Simon Newcomb, "Abstract Science in America, 1776–1876," *NA Rev.,* CXXII (1876), 96.
3. *Introductory Discourse before the Literary and Philosophical Society of New York,* May 4, 1814 (New York, 1814), p. 4. On popular indifference to science, see John C. Greene, "Science and the Public in the Age of Jefferson," *Isis,* XLIX (1958), 13–25.
4. James E. DeKay, *Anniversary Address on the Progress of the Natural Sciences in the United States: Delivered before the Lyceum of Natural History of New York, February, 1826* (New York, 1826), pp. 7–8. For a sketch of DeKay, see Appendix I.
5. Edward Hitchcock, "Remarks on Professor Eaton's Communication," *AJS* (1825), 150. For a sketch of Hitchcock, see Appendix I.
6. DeKay, *Address,* p. 70.
7. John Eberle, "A Sketch of the Improvement of Medical and Surgical Science in the United States, During the Last Thirty Years," *American Medical Recorder,* II (1819), 476, 481–82. Eberle (M.D., University of Pennsylvania, 1809) was a leading medical journalist of the early nineteenth century. For a list of the journals he edited, see fn. 30. He was also the leader in establishing the Jefferson Medical College and the Medical Department of Miami University.
8. Parker Cleaveland, *Elementary Treatise on Mineralogy and Geology* (Boston, 1816).
9. "American Geology," *Edinburgh Review,* XXX (1818), 382.
10. William Maclure, *Geology of the United States,* p. 35. Cleaveland, *Elementary Treatise,* p. 352.
11. On this point, see F. C. Gray's remarks on the Danas' claim to have discovered basalt in their survey; "Systems of Geology," *NA Rev.,* VIII (March, 1819), 413. See also DeKay, *Address,* p. 16.
12. For Hitchcock, see "The New Theory of the Earth," *NA Rev.,* XXVII (April, 1829), 260–70; for Silliman, see "Address before the Association of American Geologists and Naturalists," published in *AJS,* XLIII (1842), 217; for Eaton's adoption of fossil chronology, see "Four Cardinal Points in Stratigraphical Geology, Established by Organic Remains," *AJS,* XXX (1832), 199; or "Geological Equivalents," ibid., 132.
13. *Edinburgh Review,* XXX, 375.
14. John Locke, *An Introductory Lecture on Chemistry and Geology* (Cincinnati, 1839), p. 5. For a sketch of Locke, see Appendix I. Taking the middle way between two extremes seemed to be characteristic

of American thought in a number of areas. For one example, Emma Willard, one of three women who wrote in the major scientific journals, in an article on "Universal Terms—Disputes Concerning Them and Their Causes," *AJS*, XXIII (1833), 27, settled the question of whether general or individual terms were first invented by saying that concerning some classes generals were first and in others individuals were first.

15. William Maclure to Benjamin Silliman, December 4, 1821; printed in *AJS*, V (1822), 197. See also Isaac Lea, "A Sketch of the History of Mineralogy," *Philadelphia Journal of the Medical and Physical Sciences*, VII (1823), 288. For sketches of Maclure, Silliman, and Lea, see Appendix I.

16. Stephen G. Kurtz, "James Rush, Pioneer in American Psychology, 1786–1869," *Bull. Hist. Med.*, XXVII (1954), 50–59.

17. [Benjamin Silliman], "Remarks on Some Points of Modern Chemical Theory, With a Notice of Professor Gorham's Elements of Chemistry," *AJS*, II (1821), 330.

18. William Maclure, *Observations on the Geology of the United States of America . . . in Reference to the Accompanying Geological Map* (Philadelphia, 1817); Parker Cleaveland, *Elementary Treatise on Mineralogy and Geology* (Boston, 1816); John Gorham, *The Elements of Chemical Science* (Boston, 1819); Richard Harlan, *Fauna Americana: Being a Description of the Mammiferous Animals Inhabiting North America* (Philadelphia, 1825).

19. See Chapter II.

20. See "Scientific Societies, Aims and Activities," *AJS*, X (1826), 373–74.

21. For a table showing the growth of scientific societies in America, see Ralph S. Bates, *Scientific Societies in the United States* (New York, 1945), p. 51.

22. See Appendix II.

23. Daniel Drake, "Our Periodical," *Western Journal of the Medical and Physical Sciences* (January, 1836), 521. For a sketch of Drake, see Appendix I.

24. Daniel Drake, "Our Delay," *Western Journal of Medicine and Surgery*, I (February, 1840), 359.

25. Florian Cajori, *The Teaching and History of Mathematics in the United States* (Washington, 1890), p. 95.

26. See Appendix II.

27. *Journal of the Academy of Natural Sciences of Philadelphia*, I (1817), 1–2.

28. "Description of Four Remarkable Fishes, Taken Near the Piscataqua in New Hampshire," *Memoirs of the American Academy of Arts and Sciences*, II, Part 2 (1797), 46.

29. See Denis I. Duveen and Herbert S. Klickstein, "The Introduction of Lavoisier's Chemical Nomenclature into America," *Isis*, XLV (1954),

278, 368; and Robert Siegfried, "An Attempt in the United States to Resolve the Difference Between the Oxygen and the Phlogiston Theories," *Isis* XLVI (1955), 327.

30. John Eberle edited the *American Medical Recorder* and the *American Medical Review,* both of Philadelphia; the *Western Medical Gazette,* the *Ohio Medical Lyceum,* and the *Western Quarterly Journal of Practical Medicine,* all of Cincinnati; and the *Transylvania Journal of Medicine and the Associate Sciences,* of Lexington, Kentucky. John Bell edited the *New York Medical and Physical Journal,* the *Western Medical Reformer,* of Cincinnati; the *Select Medical Library and Eclectic Journal of Medicine,* of Philadelphia, and the *Bulletin of Medical Sciences.* Isaac Hays edited the *Philadelphia Journal of Medical and Physical Sciences* and the still existing *American Journal of Medical and Physical Sciences,* of Philadelphia; the *Cholera Bulletin,* of New York; and the *Cholera Gazette,* of Philadelphia. For Godman, see Appendix I.

31. For an account of the mathematical journals, see Florian Cajori, *The Teaching and History of Mathematics in the United States* (Bu. Ed. Circular of Info. No. 3, Washington, 1890), pp. 94 ff.

32. Benjamin Silliman, "Preface," *AJS* XVI (1829), iv–vi. For an account of Silliman's problems with the *Journal,* see his "History of the *American Journal of Science and Arts,*" in *ibid.,* L (1847), 3–18.

33. For a list of journals, see "Introduction," fn. 8.

34. Since no one has ever claimed to have actually established a single one of these generalizations, one cannot blame any specific scholar for them. Writers typically mention them in passing or seem to assume them in their work. Dirk J. Struik, for example, in *Yankee Science in the Making* (New York: 1962) mentions all of them in Chapter VII, but except for the issue of practicality they do not influence the rest of his book. Merle Curti, in *The Growth of American Thought* (New York, 1951), Chapter XIII, notes that specialization and professionalization began sometime between 1830 and 1850, but he does not elaborate upon these themes, and one gathers from his survey that virtually the only type of scientific work done in America in that period was practically oriented, natural-history investigation. For a typical statement of American emphasis on natural history, see William M. and Mabel S. C. Smallwood, *Natural History and the American Mind* (New York, 1941). A. Hunter Dupree, in what is undoubtedly the finest biography of a scientist that has yet been written, nevertheless, tends to overstate Grady's uniqueness by frequent references to the "amateurishness" and unspecialized nature of American scientists. See his *Asa Gray, 1810–1888* (Cambridge, 1959); e.g., pp. 25–26, 95, 114, 212. See also, Richard H. Shryock, "American Indifference to Basic Science during the Nineteenth Century," *Archives Internationales d'Histoire des Sciences,* No. 5 (1948), 50–65. At the 1963 meeting of the History of Science Society, where these generalizations were discussed, it was evident that

most accepted them. Of about 100 historians of science present, only John Greene, Charles Rosenberg, Nathan Reingold, and the present writer dissented from any of them.

35. Although it would be much neater to have "natural history" and "physical science" exhaust the universe of science, there were some sciences at that time that did not seem to clearly belong in either category; i.e., physiology. Under "natural history" I have included botany, zoology, mineralogy, geology, paleontology, and miscellaneous works that seem to be related to any of these; under "physical sciences" I have listed physics and chemistry, meteorology, mechanics, astronomy, and mathematics.

36. DeKay, *Address*, pp. 69 and 78.

37. Walter B. Hendrickson, "Nineteenth Century State Geological Surveys: Early Government Support of Science," *Isis*, LII (September, 1961), 370.

38. Legislative resolutions on geological surveys are summarized in George P. Merrill, *Contributions to a History of American State Geological and Natural History Surveys* (U. S. National Museum, *Bulletin* No. 109; Washington, 1920). For one effort to distinguish between the two types of interests, see Gerald D. Nash, "The Conflict between Pure and Applied Science in Nineteenth Century Public Policy: The California State Geological Survey, 1860–1874," *Isis*, LIV (June, 1963), 217.

39. Hitchcock built the largest collection in the world of such tracks (now known to have been dinosaur tracks) at Amherst College and wrote numerous articles describing his researches. See, for example, "Report on Ichnolithology, or Fossil Footmarks," *AJS*, XLVII (1844), 292, which contains a history of the discoveries. The first suggestion that these were reptilian was Roswell Field's "Ornithichnites, or Tracks Resembling Those of Birds," *AAAS Proc.*, 1859, 337–40; *AJS*, XXIX (1860), 361. This was demonstrated in 1893 when Marsh dug one up in the West.

40. James Hall to A. D. Bache, January 14, 1852. Rhees Collection, Huntington Library.

41. Dupree, *Science in the Federal Government*, Chapters I–II.

42. I am indebted to A. Hunter Dupree, *Ibid.*, p. 26 for this story. A part of the reason that historians have so harshly judged American expeditions can be attributed to their misunderstanding of the motivations behind European expeditions of a similar nature. For example, the Laperouse expedition which departed from France to circumnavigate the earth in 1785 has generally been considered a purely scientific expedition. But Seymour L. Chapin in *Astronomy and the Paris Academy of Sciences During the Eighteenth Century* (Ph.D. Thesis, UCLA, 1964), 204–206, has argued convincingly that in reality its motives were political—simply another chapter in a continuing competition for empire between France and England. The decision to make scientific use of the expedition came almost as an afterthought.

43. Jefferson explained how Congress in secret proceedings had "yielded to a proposition" in a letter to Benjamin Smith Barton, Feb. 27, 1803 (Barton Correspondence, Historical Society of Pennsylvania). The purpose of the letter was to ask Barton to instruct Lewis in botany. For Jefferson's belief in the fullness of the creation, see Daniel J. Boorstin, *The Lost World of Thomas Jefferson* (New York, 1948), pp. 24, 36–40.

44. For a brief contemporary account of the Long expedition, see Edward Everett, "Review of Edwin James (comp.) Account of the Long Expedition," *NA Rev.*, XVI (1823), 242. James' compilation of notes taken by naturalists on the expedition was published in two volumes in Philadelphia, 1822–1823.

45. John Torrey, "Collection of Plants Made During A Journey to the Rocky Mountains," *Annals of the New York Lyceum of Natural History*, II (1828), 161.

46. Samuel Jackson, "On the Condition of the Medicines of the United States—and the Means of Their Reform: An Introductory Lecture Delivered in the Philadelphia College of Pharmacy," *Philadelphia Journal of the Medical and Physical Sciences*, V (1823), 212.

47. Benjamin Silliman, "Notice of Thomas Thomson's Chemistry of Organic Bodies and Vegetables," *AJS*, XXXVI (1839), 203. E. B. Hunt, "Views and Suggestions on the Practice and Theory of Scientific Publication," *AJS*, XXVI (1858), 25.

48. More liberality in defining "related areas" would have resulted in an even more impressive percentage; but I have confined the meaning to work actually contributing to the major interest of the author. For example, zoology, which describes extant animals, and paleontology, which describes extinct animals, are clearly related, as are descriptive mineralogy and analytical chemistry. But zoology and botany are not related.

49. Charles U. Shepard, "Reply to a Notice of Shepard's Mineralogy," *AJS*, XLVIII (1845), 169. The review referred to is in *AJS*, XLVII (1844), 333. For a sketch of Shepard, see Appendix I. For a similar definition, see [Constantine Rafinesque], "Survey of the Progress and Actual State of Natural Sciences in the United States of America, From the Beginning of this Century to the Present Time," *American Monthly Magazine and Critical Review*, II (1817), 81. In spite of the comments made above, I have followed usage in classifying geology as natural history in the tables.

50. The seven men were John Bachman, clergyman; Charles L. Bonaparte, nephew of Napoleon and a political refugee; N. M. Hentz, a professor of modern languages; Isaac Lea, publisher; E. C. Herrick, Yale University librarian; and William Maclure and Henry Seybert, independently wealthy gentlemen. Only Herrick, an astronomer, and Seybert, a chemist, worked outside natural history areas; and even though Seybert was not employed, he was a graduate of the Paris School of Mines. It is significant that there is only one practicing clergyman on

the list. While religion may very well have dominated science during the period, it was not because clergymen were scientists. Robert Siegfried, in *A Study of Chemical Research Publications from the United States before 1880* (Ph.D. Thesis, U. of Wisc., 1952), p. 139, found that only five of the fifty-seven most productive chemists in America before 1821 never had anything to do with teaching.

51. The most unfortunate thing about this error is that it is self-perpetuating. Phyllis A. Richmond, in her very useful compilation of *American Fundamental Research in the Nineteenth Century* (unpublished Master's paper, Western Reserve University School of Library Science, 1956), found 18 contributors to the eminently practical medicine, 33 to the natural-history sciences, but only 2 to physiology and 6 to the physical sciences. Mrs. Richmond therefore rightly concluded that American scientific research interests had been grossly underestimated, but she wrongly concluded that there was little research interest in the physical sciences. The reason for this conclusion, of course, was that the only available authorities had concentrated on medicine and natural history, because they knew that Americans had made niches for themselves in the history of the progress of these sciences. Her list, for example, does not include either C. T. Jackson, Robert Hare, or C. J. Page. For sketches of these men and their work, see Appendix I.

52. Scholars who have attempted to compare English and American institutional supports for science have generally based their conclusions on an inadequate understanding of English institutions. For a disenchanted account of English institutions during that period, see the opening chapters of D. S. L. Cardwell, *The Organisation of Science in England: A Retrospect* (London, 1957).

II. The Scientific Profession

1. The chief authority for this statement is Theodore Hornberger, *Scientific Thought in American Colleges, 1638–1800* (Austin, 1945), whose count is based upon Samuel Miller's *A Brief Retrospect of the Eighteenth Century* (New York, 1803). A persistent search on my own part has failed to disclose any more.

2. *The Writings of Thomas Jefferson*, ed. Andrew A. Lipscount (Washington, 1905), XIX, 456.

3. For a brief account of academic scientists in the late eighteenth century, see Brooke Hindle, *The Pursuit of Science in Revolutionary America, 1735–1789* (Chapel Hill, 1956), Chapter V.

4. By 1800 there had existed twenty-seven colleges in the United States, many of them short-lived. Sixteen more were founded by 1820; and eighty-nine new ones began operations between 1821 and 1850. These figures are my own count, based upon a variety of sources. For one example of the growth of enrollment in medical schools, which tended to set the standards for scientific instruction, see Robley Dung-

lison's table of enrollments in the Jefferson Medical College, where he taught during the 1840s. In 1840, there were 163 students; by 1846, the number had grown to 493. *The Biographical Ana of Robley Dunglison*, ed. F. X. Radbill (American Philosophical Society, 1964), p. 94. Since each professor received a standard sum from each student, it is obvious that there were healthy increases of income during those seven years. On the importance of decentralization and academic competition, see Joseph Ben-David, "Scientific Productivity and Academic Organization in Nineteenth Century Medicine," *American Sociological Review*, XXV (December, 1960), 828–43.

5. John Ware, "Gorham's Chemistry," *NA Rev.*, IX (1819), 134.

6. Quoted in George P. Fisher, *Life of Benjamin Silliman* (New York, 1866), I, 289.

7. While it is certainly true that only a small group of men in the previous century had the mathematical ability to work with Newton's calculus, Newton's system still appeared understandable to all who commented on it. At any rate, it caused no such bewilderment as those one finds in the nineteenth century.

8. For Dewey's statement, dated March 14, 1842, see Dupree, *Asa Gray*, p. 102. For Jefferson, see his letter to Dr. John Manners, February 22, 1814. Quoted in full in Edmund H. Fulling, "Thomas Jefferson, His Interest in Plant Life as Revealed in His Writings, II," *Bulletin of the Torrey Botanical Club*, 72, No. 3 (May, 1945), 250.

9. London, 1853, pp. vii–x.

10. "Lindley's Natural System of Botany," *AJS*, XXXII (1837), 300. But see his much more candid statement of the complexity of the natural system in "Henfrey's Botany," *AJS*, 2d ser., XXIV, 434.

11. [Dr. Ray], "DeCandolle's Botany," *NA Rev.*, XXXVIII, 1834, 55.

11a. For the absence of the appeal in the first quarter of the century, see John C. Greene, "Science and the Public in the Age of Jefferson," *Isis*, XLIX (1958), 25.

12. "The Wonders of Science Compared with the Wonders of Romance," in *Religious Truth, Illustrated from Science*, Boston, 1857, pp. 139–40.

13. VI, 1802, 434.

14. "Notes to the American Edition of Henry's Chemistry," in William Henry, *An Epitome of Experimental Chemistry*, 2nd American Edition from the 5th English Edition (Boston, 1810), ii. Quoted in John C. Greene, "Science and the Public in the Age of Jefferson," Isis, XLIX (1958), 24.

15. James E. DeKay, *Anniversary Address on the Progress of the Natural Sciences in the United States;* delivered before the *Lyceum of Natural History of New York*, February, 1826 (New York, 1826), p. 8.

16. [F. C. Gray], "American Forest Trees," *NA Rev.*, XLIV, April, 1837, 361.

17. [Nathaniel Chapman], "Prospectus," *Philadelphia Journal of the*

Medical and Physical Sciences, I (1820), x. Chapman (1780–1853) received his medical degree under Benjamin Rush at the University of Pennsylvania in 1801, and studied for three years in Edinburgh. Other than his contributions to medical journalism, Chapman is remembered as the founder, in 1817, of the medical institute of Philadelphia, which may be considered the first postgraduate school in the United States. He was elected the first president of the American Medical Association, by acclamation, in 1847; and he served as president of the American Philosophical Society from 1846 to 1848. The same kind of statement as that made by Chapman is in the "Prospectus" to G. W. Featherstonehaugh's *Monthly American Journal of Geology and Natural Science,* I (1831). For a study of Chapman's journal, see Charles L. Losano, "The Philadelphia Journal of the Medical and Physical Sciences, 1820–1827," *Bulletin of the History of Medicine,* XXXIV (1960), 75–79.

18. [Nathaniel Chapman], "American Medicine," *Philadelphia Journal of the Medical and Physical Sciences,* IX (1824), 407, 408.

19. [J. G. Cogswell], "Dr. Webster on St. Michael and the Azores," *NA Rev.,* XIV (1822), 50. Cogswell (1786–1871) studied law under Fisher Ames, attended Göttingen, and tutored at Cambridge. At the time he wrote the review he was professor of geology and mineralogy at Harvard University. The work he referred to was *A Description of the Island of St. Michael, Comprising an Account of Its Geological Structure; With Remarks on the Other Azores or Western Islands* (Boston, 1821). For a sketch of Webster see Appendix I.

20. [J. D. Godman], "Review of Major Long's Second Expedition," *NA Rev.,* XXI (1825), 178, 179. For a sketch of Godman see Appendix I.

21. "Redfield's and Espy's Theories," *New York Review,* VII (April, 1840), 300. For a sketch of Redfield see Appendix I.

22. "Astronomy," *American Quarterly Review,* III (June, 1828), 319.

23. American State Papers, *Misc.* II, 753. Quoted in Dupree, *Science in the Federal Government,* p. 34. Around the mid-century there was a great deal of discussion of this issue at meetings of the American Association for the Advancement of Science. See, for example, *Proceedings,* II (1849), 381–83; IV (1851), 155–57.

24. C. O. Paullin, "Early Movements for a National Observatory, 1802–1842," *Records of the Columbia Historical Society,* XXV (1923), 42–43. Richard Rathbun, "The Columbian Institute for the Promotion of Arts and Sciences," U. S. National Museum, *Bulletin,* No. 101 (Washington, 1917), 64.

25. "National Institute," *American Whig Review,* II (1845), 238.

26. *AJS,* XVI (1829), note, p. 225.

27. Even though A. Hunter Dupree, in *Science in the Federal Government,* p. 114, concludes that by 1860 government activities (including that of states) were "possibly" the largest and most important source of funds and employment for science in the country, one can still gather from his work that the support was hardly adequate.

28. *The Medical Repository,* II (1815), 259–60. Quoted in John C. Greene, "Science and the Public," p. 24.

29. "DeCandolles Botany," *NA Rev.,* XXXVIII (1834), 33.

30. *Ibid.* 31. "Wonders of Science," p. 187.

32. "The Study of Natural History," *Knickerbocker,* XXV (1845), 292.

33. "Dr. Webster's Manual of Chemistry," *NA Rev.,* XXIII (1826), 350.

34. "The Study of Natural History," *NA Rev.,* XLI (1835), 428. This conservative appeal for widespread education has been studied by Rush Welter in *Popular Education and Democratic Thought in America* (New York, 1962).

35. [Timothy Walker], "Defense of Mechanical Philosophy," *NA Rev.,* XXXIII (1831), 123.

36. Samuel Tyler, "The Influence of the Baconian Philosophy," *Princeton Review,* XV (1843), 483.

37. D. March, "Physical Science and the Useful Arts in Their Relation to Christian Civilization," *New Englander,* IX, 1851, 492.

38. "The Relations and Mutual Duties Between the Philosopher and the Theologian," *Bibliotheca Sacra,* X (1853), 191–92, 194. Reprinted in *Religious Truth,* 54–97.

39. J. D. Dana, "Science and Scientific Schools," *Am. J. Educ.,* II, 1856, 363. A Commencement Address delivered at Yale College, August, 1856.

40. "Observations upon certain passages in Mr. Jefferson's Notes on Virginia which appear to have a tendency to subvert religion and establish a false philosophy," pamphlet, 1804 (in the collection of the Newberry Library, Chicago).

41. J. Haven, Jr., "Natural Theology," *Bibliotheca Sacra,* VI (1849), 614.

42. Benjamin Peirce, "Presidential Address," *AAAS Proc.,* VIII (1854), 14.

43. E. S. Gannett, "The Value of Natural Religion," *Christian Examiner,* XXXIV (July, 1843), 296. Rev. L. F. Smith, "Lectures on the Evidences of Christianity" (Review of Hopkins' Lowell Lectures), *Christian Review,* II (July, 1846), 229.

44. James Marsh, Introduction to Samuel Taylor Coleridge, *Aids to Reflection,* in *The Complete Works of Samuel Taylor Coleridge,* ed. W. G. T. Shedd, 7 Vols (New York, 1884), I, 100.

45. *Six Days of Creation* (Schenectady, 1855), p. 397. See also pp. 404–407.

46. "Science and the Bible," *Bibliotheca Sacra,* XIII (January, 1856), 81.

47. "Science and the Bible, Number II," XIII (July, 1856), 632.

48. *NA Rev.,* LX (1845), 426–78.

49. "Vestiges of Creation and Its Reviewers," *New Englander,* IV (1846), 115.

50. H. D. Rogers to W. B. Rogers, Jan. 24, 1845. See also John F. Frazier to S. S. Haldeman, Jan. 31, 1846; A. A. Gould to Haldeman, Jan. 28, 1840; Haldeman Papers, ANSP.

51. *NA Rev.*, LXIII (1846), 505–506, quoted from *Explanations*, 124–25.

52. *Ibid.*, p. 506.   53. *Ibid.*, p. 504.

54. Asa Gray to J. D. Hooker, July 13, 1858.

55. "Introduction" to the American Edition of Hugh Miller, *The Footprints of the Creator; Or, the Asterolepsis of Stromness* (New York, n.d., introduction dated 1858), p. xxxvi.

56. *Ibid.*, p. 244.   57. "Preface," *AJS*, XVI (1829), iv.

58. Silliman's difficulties with his fellow scientists came later, when a new generation came to leadership in the scientific community. In his later years he was often accused of falling down on the job of censorship, as Dana said, allowing his "excessive benevolent feelings" to get the better of his good sense. According to Dana, writing in 1851, an author, pleading for admission to the journal "at once enlists his sympathies" and Silliman would publish the article "even though hundreds are dissatisfied with it." James D. Dana to Alexander Dallas Bache, September 6, 1851 (Huntington Library Collection). On Silliman's retirement from active editorship Dana and Silliman's son took over the job of acting as watchdogs for the profession.

59. Asa Gray, "Notice of the Botanical Writings of the Late C. S. Rafinesque," *AJS*, LX (April, 1841), 221, 239, 241.

60. The coolness of American Catholics and High Church Episcopalians, amounting sometimes to hostility, is a particularly interesting exception that will be dealt with in a later chapter.

61. Thomas Carlyle, "Signs of the Times," *Edinburgh Review*, No. 98 (June, 1829), 454.

62. Walker, *NA Rev.*, XXXIII, 133, 123.

III. The Reign of Bacon in America

1. [Edward Everett], "Character of Lord Bacon," *North American Review*, XVI (1823), 200. Italics are in the original. The work being reviewed was Mallet's *Life of Bacon*.

2. *Dissertation: Exhibiting the Progress of Metaphysical, Ethical, and Political Philosophy Since the Revival of Letters in Europe;* ed. Sir William Hamilton (Edinburgh, 1854), p. 64. The first edition was published in 1815.

3. See, for example, "Review of the Principia of Newton," XI (1826), 240. James R. Newman, in his essay on Bacon in *Science and Sensibility* (New York, 1961), Vol. 1, 90, traces to Voltaire the myth that Bacon anticipated Newton in the formulation of the law of gravitation. James C. Crowther, the latest Bacon fan, claims only that Bacon "formulated

the great problem to which Isaac Newton gave the answer," *Francis Bacon, The First Statesman of Science* (London, 1960), p. 67. He does, however, claim that Bacon influenced Newton's work on colors, was "groping toward photography and the cinema," and formulated the basic principle of information theory (pp. 35, 80). Critical treatments of Bacon are, unfortunately, difficult to find in English, but there is one excellent work: Fulton H. Anderson, *The Philosophy of Francis Bacon* (Chicago, 1948). I am indebted to this work for most of my comments on Bacon himself. Although English writers have not been particularly interested in the origins of Baconianism, there is an excellent work in Italian, Paolo Rossi, *Francesco Bacone, Dalla Magia Alla Scienza* (Bari, 1954), and there is an older work in German, Walter Frost, *Bacon und die Naturphilosophie* (München, 1927), which explores Bacon's relation to medieval logic, alchemy, and to such predecessors as Copernicus and Kepler. Although she is an admirer of Bacon, Dorothea Krook, in "Two Baconians: Robert Boyle and Joseph Glanvill," *Huntington Library Quarterly*, XVIII (May, 1955), makes a careful distinction between the philosophy of Bacon and "Baconianism." One of her subjects, Glanvill, would have been an excellent example of nineteenth-century Baconianism.

4. John Esten Cooke, *Treatise on Pathology and Therapeutics* (Lexington, 1828), Vol. I, preface. Cooke (M.D., University of Pennsylvania, 1805) was at the time he wrote professor of the theory and practice of medicine at Transylvania University and co-editor of the *Transylvania Journal of Medicine and the Associate Sciences*. He, along with Charles Caldwell, was probably the most potent factor in shaping medical thought in the Southwest. See Lunsford P. Yandell, "The Life and Writings of John Esten Cooke," *American Practitioner*, VI (July, 1875), 1 ff.

5. Quoted in John C. Greene, *The Death of Adam* (Ames, Iowa, 1959), p. 171. Greene makes the point that this is derived from Cuvier's identification of the conditions of existence with final causes.

6. John C. Warren, "New Theory of Human Deformity," *Boston Medical and Surgical Journal*, III (Tuesday, November 9, 1830), 642. John Collins Warren (M.D., Harvard, 1797; Edinburgh, 1800) was professor of anatomy and surgery at Harvard and co-editor of the *Boston Medical and Surgical Journal*. See, Edward Warren, *Life of John Collins Warren* (Boston, 1860).

7. George J. Chace, "Of the Divine Agency in the Production of Material Phenomena," *Bibliotheca Sacra*, V (May, 1848), 347. Chace was professor of chemistry and geology at Brown University. See also [Francis Bowen] "Chalmers' Theology," *NA Rev.*, LIV (October, 1842), 356 ff; or Samuel Stanhope Smith, *The Lectures, Corrected and Improved, Which Have Been Delivered for a Series of Years in the College of New Jersey; on the Subjects of Moral and Political Philosophy*, 2 vols. (Trenton, New Jersey, 1812), I, 20.

8. Dugald Stewart, *Progress of Philosophy*, p. 64.

9. "Bacon's Philosophy," *Methodist Quarterly Review*, XXXI (January, 1847), 22. Although "philosophy" was often used as the equivalent of "science" in the early nineteenth century, the former word was beginning to acquire the more modern meaning, and this resulted in a great deal of ambiguity in the use of the term—an ambiguity that makes it impossible to understand what was meant without a context. Generally, the most correct translation for "philosophy" would be "those principles of reasoning and fundamental assumptions upon which scientific investigations are carried out." Except in direct quotations, where I think the meaning will be obvious, I will substitute the most appropriate modern terminology for each case.

10. Sir David Brewster, *Life of Sir Isaac Newton* (London, 1831), p. 297.

11. Thomas Babington Macaulay, *Critical and Miscellaneous Essays* (Philadelphia, 1842), Vol. II, 474. See also, Macaulay's article "Lord Bacon" in the *Edinburgh Review*, LXV (1837), 1–104.

12. Joseph De Maistre, *Examen de la Philosophie de Bacon* (Lyon, 1845), especially Tome II, p. 313. This was a posthumous publication, a different edition having been published at Paris in 1836. As far I could determine, 1836 was the first time any of this material had been published, although De Maistre died in 1821.

13. "Brande's Encyclopedia," *United States Catholic Magazine and Monthly Review*, III (January, 1844), 3.

14. "Utility of Physical Sciences," *The Catholic Expositor and Literary Magazine*, II (1842), 251.

15. In all the scientific literature through the mid-1840s I have found no more than occasional passing references to Transcendentalism. Sometimes (see below) it was used as an all-inclusive term for any apriorism. It did not become recognized as a major threat to science until after Louis Agassiz added the weight of his scientific reputation to Transcendental philosophy. For the conflict between Agassiz and the more empirical-minded scientists, see Dupree, *Asa Gray*, Chapter XII.

16. It is a commonplace observation that the classic formulation of a system often comes just at the time that system has outlived its usefulness. E. B. Titchener's *Experimental Psychology* is a case in point.

17. For the comments on Tyler see *Princeton Review*, XIX (1847), 125; *Southern Presbyterian Review*, January 1859, p. 676; Francis Lieber, *On Civil Liberty and Self Government* (1859); 199n.; E. A. and G. L. Duyckinck, *Cyclopedia of American Literature* (New York, 1856), II, 519; Charles Lanman, *Haphazard Personalities Chiefly of Noted Americans* (Boston, 1886), pp. 304–15. The fact that the Library of Congress does not possess a copy of the first edition of Tyler's *Discourse on the Baconian Philosophy*, the work that his contemporaries bestowed such lavish praise upon, is a rough indication of how completely he has been forgotten by history. The longest biographical account of Tyler I could

find was that of Charles Lanman, pp. 304–15. Lanman was a personal acquaintance of Tyler who had been commissioned by the Tyler family to write a biography, but for some reason he never wrote more than this brief sketch.

18. To call an idea "metaphysical" was a common way to denounce it. J. B. Stallo's comment will give some indication of the general dislike of the metaphysical or speculative: "Unfortunately, the materialistic, utilitarian tendencies, which at present pervade every branch of science under color of a misconceived Baconism, have revoked every alliance between philosophical pursuits and the investigation of nature. Speculation is perfectly disavowed (sometimes justly, perhaps), and the very worst passport which a naturalist could carry about him is that of a metaphysician." *General Principles of the Philosophy of Nature with an Outline of Some of Its Recent Developments Among the Germans, Embracing the Philosophical Systems of Schelling and Hegel and Oken's System of Nature* (Boston, 1848), p. vii. Stallo was professor of analytical mathematics, natural philosophy, and chemistry at St. John's College, New York.

19. "On the Baconian Philosophy," *Princeton Review*, XII (July, 1840), 350. The work which was used as the occasion for the article was Basil Montague, *The Works of Lord Bacon* (London, 1838). Tyler's *Discourse on the Baconian Philosophy* (Baltimore, 1844) contains nothing that is not found in the articles, and I, therefore, prefer to use the latter. All but one of the articles were written before the book, and since they were written as reviews they more clearly illustrate his relationship to the other intellectual currents of the time. Publishing seriously intended works on the philosophy of science in theological reviews was not at all uncommon during that period; Edward Hitchcock earlier, and James Dwight Dana later, published some of their most significant works in such journals.

20. *Ibid.*, p. 364. In arguing that there was no conflict beween syllogism and induction, Tyler was not mirroring the beliefs of his practicing scientist contemporaries, undoubtedly because he possessed a bit more philosophical sophistication than they. The usual belief was that induction and the syllogism were rival modes of reasoning and the syllogism was of absolutely no value, or at the most, of very limited value. In this case, Tyler was more Baconian than they. His contemporaries in the philosophical world were divided on the issue; Richard Whately and William Whewell agreed with Tyler, while Cousin and John Stuart Mill disagreed. Mill's viewpoint on scientific reasoning was that it proceeded from particular to particular by simple substitution; Whewell was one of the earliest to argue that scientific reasoning was hypothetico-deductive.

21. A few of the more perceptive observers did speak out against this tendency to equate science with classification. See, for example,

John D. Godman, *American Natural History* (Philadelphia, 3rd ed., 1846), p. xiii. The first edition was published in 1823.

22. *Ibid.*, p. 363. See also "Whately's Logic," pp. 297–98.

23. [George B. Emerson], "Notice of De Candolle," *AJS*, XLII (1842), 217. The outstanding example of a man who sought to reduce science to a taxonomy was Lavoisier, whose method has been brilliantly analyzed by Charles C. Gillispie in *The Edge of Objectivity: An Essay in the History of Scientific Ideas* (Princeton, 1960), pp. 236 ff.

24. *Ibid.*, pp. 369–70. Apparently Tyler no longer thought it necessary to avoid the expression "inductive evidence," as he had in 1837. The expression, he had then thought, implied that induction was a species of evidence, which it was not. He then preferred the expression "evidence on which the inductive process [a species of investigation] is conducted." Whately's Logic," pp. 304–305. In his use, however, the two expressions mean the same thing.

25. Samuel Tyler, "Psychology," *Princeton Review*, XV (April, 1843), 227. According to Merle Curti, Locke remained America's philosopher through the first half of the nineteenth century, and he rejects the notion, which he alleges to have been common among scholars, that Locke was replaced by the Scottish philosophers during this period. However, I do not think that he adequately considered the degree of "reinterpretation" that was made of Locke—primarily by Dugald Stewart. For Curti's interpretation, see "The Great Mr. Locke: America's Philosopher, 1783–1861," *Huntington Library Bulletin*, No. 11 (April, 1937), 107–52. For Stewart's interpretation of Locke, see his *Progress of Philosophy*, pp. 220–37.

26. *Ibid.*, pp. 249–50.

27. *Ibid.*, p. 242. For a good example of the argument from inconceivability, see Robert Hare, "A Letter to Prof. Faraday, on Certain Theoretical Opinions," *AJS*, XXXVIII (1840), 1 ff. William Whewell, however, took exception to the test of conceivability, and turned the common argument around. That which was a necessary consequence of a truth was true and *its* negation inconceivable, even though "to common minds," it might not appear to be the case. See his "Demonstration That All Matter Is Heavy," *AJS*, XLII (1842), 266.

28. *Ibid.*, p. 245. Even Elisha Bartlett, the most thorough-going empiricist in America at the time, adhered to Reid's justification for causation. Quoting Whewell, Bartlett said that it was sufficient for him that the idea of cause was "an indestructible conviction, belonging to man's speculative nature." *An Essay on the Philosophy of Medical Science* (Philadelphia, 1844), p. 26. For more on Bartlett, see Chapter vii.

29. Tyler, "Psychology," pp. 247, 250.

30. [Samuel Tyler], "Influence of the Baconian Philosophy," *Biblical Repertory and Princeton Review*, XV (October, 1843), 483.

31. *Ibid.*, pp. 484–93, 506.

32. *Ibid.*, p. 494. In a later work he credited Sir William Hamilton

with having saved the Baconian philosophy from being the "dirt philosophy" into which some of Bacon's "heretical disciples" had attempted to transform it; "Sir William Hamilton and His Philosophy," *Princeton Review*, XVII (1855), 599.

33. *Ibid.*, pp. 495–97. By "disinterested" Tyler meant "directed toward no particular object."

34. *Ibid.*, pp. 498–99, 504–505.

35. [Samuel Tyler], "The Connection Between Philosophy and Revelation," *Princeton Review*, XVII (July, 1845), 382. That is to say, he did not admit innate ideas in the Cartesian sense (see above). Scottish philosophers always insisted upon a distinction between innate ideas and ideas derived from "conciousness." The latter they regarded as an empirical procedure. For the charge that apriorism entails atheism, see also [Francis Bowen] "Review of Chalmers' Theology," *NA Rev.*, LIV (Oct., 1842), 360.

36. *Ibid.*, pp. 402–403.

37. *Ibid.*, p. 387. With certain qualifications, one must grant that Tyler did hold to his rule in theological controversy. Although he did usually include arguments from nature, the greater part of his criticisms of religious works was scriptural. In two works in the *Princeton Review*, he revealed himself as a highly competent Greek scholar and philologist. He was neither a narrow literalist nor was he willing to accept purely allegorical interpretations. His guiding principle seemed to be that words are never, not even in the Bible, an accurate representation of reality, and the correct form of exegesis is therefore to determine exactly what the writer meant by the words he was using—roughly a cultural anthropological approach. See, "Balfour's Inquiry," VIII (1836), 327 ff. and "Bush on the Soul," XVIII (1846), 219 ff.

38. *Ibid.*, p. 405. Writing a few years later, an American reviewer of Chalmers who claimed to follow the Baconian method in his systematic theological writings, confessed his surprise that the necessity of following that method had not been more universally acknowledged. See W. I. Budington, "Chalmers on the Inductive Method in Theology and the Nature of Christian Doctrine," *New Englander*, VIII (May, 1850), 203.

39. "Psychology," 249–50. "Psychology," [Review of Frederick A. Rauch's Psychology], *Baltimore Literary and Religious Magazine*, VII (August, 1841), 349.

40. For an example of Tyler's anti-Catholicism, see his opening comments to the review of Stone's book, "Connection between Philosophy and Revelation," p. 381.

41. *Ibid.*, p. 408. "Evangelical," as Tyler used the term, should be understood as indicating those forms of Protestant Christianity which denied that revelation could be submitted to the test of reason. It would include such groups as the Methodists, Presbyterians, Baptists, and some Congregationalists, while definitely excluding Unitarians, Universalists,

High Church Episcopalians and, of course, Roman Catholics. Following Tyler, I also use the term in this broad sense, which is not exactly like the current technical usage, where it implies an unusually intense concern with conversion as a concrete experience.

42. Hugh D. Evans, "Notices of Books," *The True Catholic: Reformed, Protestant and Free,* II (June, 1844), 124–25. Evans' definition of rationalism is from his article, "Rationalism," *True Catholic,* II (November, 1844), 385. Evans edited the journal under the auspices of the Bishop of Baltimore. It is obvious from his comments that he never understood the nineteenth-century subtleties that had been read into Lockean psychology.

43. "The Protestant Principle," *New Englander,* II (1844), 66.

44. Bishop Doane of New Jersey, Bishop Brownell of Connecticut, and Bishop Whittingham of Maryland were leaders in the movement. The *New York Review* was also highly favorable to Oxford Theology. On this point, see the article, "Oxford Theology," VI (January, 1840), 198–238.

45. T. E. Bond, "What is Methodism?" *Methodist Quarterly Review,* XXX (October, 1848), 487.

46. "Brande's Encyclopedia," p. 3.

47. Even the editor of an English journal, in 1844, would grant Americans superiority to the British in two respects—in advanced theological studies and in providing theological information to the people in popular versions. "American Theology," copied from *Chambers' Journal* in *Littell's Living Age,* II (1844), 57–58. Needless to say, both the advanced studies and the popular information were Protestant.

IV. The Philosophy in Action

1. For a sketch of Loomis, see Appendix I.

2. For a sketch of early efforts to determine longitude by means of the magnetic compass, see Seymour L. Chapin, "A Survey of the Efforts to Determine Longitude at Sea, 1660–1760, Part I: Variation of the Magnetic Compass," *Navigation,* III (1952), 188–91.

3. "Observations on the Variations of the Magnetic Needle, Made at Yale College in 1834 and 1835," *AJS,* XXX (1836), 221.

4. On the Variation and Dip of the Magnetic Needle in Different Parts of the United States," *AJS,* XXXIV (1838), 306.

5. See for example, *ibid.,* 290 ff., and "On the Variation and Dip of the Magnetic Needle in the United States," *AJS,* XXXIX (1840), 41 ff. Alexander Dallas Bache had already used the method of large scale coordination of data in 1834. At his request, the Secretary of War had sent out circulars to military posts asking whether any unusual meteoric display occurred on the night of November 13, 1834. For Bache's report, which only established that the meteoric shower at New Haven was a

local occurrence, see "Replies to a Circular in Relation to the Occurrence of an Unusual Meteoric Display on the 13th Nov., 1834," *AJS,* XXVIII (1835), 305. This is an informal type of aid to science in the co-ordination of data that should be studied in greater detail than has been done.

6. See above, fn. 5, also, "Observations on the Magnetic Dip in the United States," *APS, Trans.,* VIII (1845), 285 ff.

7. John H. Lathrop, "On the Connexion Between the Theory of the Earth and the Secular Variations of the Magnetic Needle," *AJS,* XXXVIII (1840), 68 ff. Lathrop (Yale, 1819; LL.D. Hamilton College, 1845) had previously been professor of mathematics and natural philosophy at Hamilton College, but at the time of writing the article was professor of law, history, political economy, and civil polity in the same institution. See, *Yale College Obituary Record* for 1866.

8. F. C. Gray, "Systems of Geology," *NA Rev.,* VIII (March, 1819), 405. Francis Calley Gray (Harvard, 1809; LL.D., Harvard, 1841) was an independent scholar and writer.

9. The remark was made by James D. Dana in a letter to A. D. Bache dated September 6, 1851 (Huntington Library collection).

10. William C. Redfield, "Remarks on the Prevailing Storms of the Atlantic Coast of the North American States, *AJS, XX* (1831), 17 ff. For biography, see Appendix I.

11. James P. Espy, "Deductions from Observations Made, and Facts Collected on the Path of the Brunswick Spout of June 19th, 1835," *APS, Trans.,* n.s., V (1837), 421 ff.

12. *Philosophical Magazine,* 3rd series, XVIII (1841), 515.

13. "Instructions to the Officers of the Antarctic Expedition," *Philosophical Magazine,* 3rd series, XV (1839), 189 ff.

14. The reception by the French Academy of Espy's paper "amounted to applause"; G. G. Davis, "Redfield, Reid, Espy and Loomis on the Theory of Storms," *NA Rev.,* LVIII (1844), 361.

15. "On the Storm Which Was Experienced Throughout the United States about the 20th of December, 1838," *APS, Trans.,* VII (1841), 160.

16. *Ibid.,* 160, 161.

17. *Ibid.,* 161. European storms, unlike West Indian hurricanes, are immediately followed by a cold front (the term "front" did not even come into existence until after World War I) which causes the relationship Loomis noticed. Local peculiarities in the United States often cause a similar deviant relationship. For example, the Santa Ana Winds, flowing out of the Mojave Desert of California, cause extremely hot temperatures in the Los Angeles Basin, accompanied by a rising barometer. The reason for this is that the winds blow downhill and the air is therefore compressed, causing a rise in pressure.

18. *Ibid.,* 162.

19. Redfield, in commenting on Loomis' observations, concluded that the axis of the 1836 storm must have passed westward and northward of his limits of correct observation. This, he suggested, was why Loomis

was unable to establish its whirlwind character. "Observations on the Storm of December 15, 1839," *AJS*, XLII (1842), 116.

20. "On the Tornado Which Passed Over Mayfield, Ohio . . . With Some Notices of Other Tornadoes," *AJS*, XLIII (1842), 278.

20a. William C. Redfield, "Reply to Dr. Hare's Objections to the Whirlwind Theory of Storms," *AJS*, XLII (1842), 308. Bache and Olmsted had also diagrammed fallen trees crossing each other. See, *Journal of the Franklin Institute*, III (1841), 273, 276; *AJS*, XXXIII (1838), 369.

21. "On Two Storms Which Were Experienced Throughout the United States in the Month of February, 1842," *APS, Trans.*, IX (1846), 161 ff.

22. The circular is quoted by H. A. Newton in his sketch of Loomis in National Academy of Science, *Biographical Memoirs*, III (1895), 225.

23. "On the Tornado Which Passed Over Mayfield, Ohio . . ." *AJS*, XLIII (1842), 278.

24. H. A. Newton, pp. 222–23, 236.

25. "Two Storms . . . in the Month of February, 1842," *APS, Trans.*, IX (1846), 183–84.

V. A Deluge of Facts

1. On the random nature of early fact gathering, see T. S. Kuhn, *The Structure of Scientific Revolutions* (Chicago, 1962), p. 15.

2. John D. Godman, *American Natural History* (Philadelphia, 3rd ed., 1846), p. xii. See also [Dr. Ray] "De Candolle's Botany," *NA Rev.*, XXXVIII (1834), 36.

3. Roswell Park, *Pantology; or, A Systematic Survey of Human Knowledge* (Philadelphia, 1841).

4. For the quotes and for a discussion of Lavoisier's classification system, see Douglas McKie, *Antoine Lavoisier* (New York, 1952), Chapter XXII.

5. Robert Hare, "Some Encomiums upon the Excellent Treatise of Chemistry by Berzelius; also Objections to his Nomenclature," *AJS*, XXVII (1835), 70.

6. John Ware, "Gorham's Chemistry," *NA Rev.*, IX (1819), 113–15.

7. Benjamin Silliman, "Notice of Thomas Thomson's Chemistry of Organic Bodies and Vegetables," *AJS*, XXXVI (1839), 203.

8. Denison Olmsted, "On the Present State of Chemical Science," *AJS*, XI (1826), 352.

9. Philosaphes, "Philosophy of Medical Prescriptions," *Boston Journal of Medicine and Surgery*, III, No. 29 (Tuesday, August 31, 1830), 458, 457. For the emphasis on nosology, see Richard H. Shryock, *Medicine and Society in America, 1660–1860* (New York, 1960), p. 65.

10. William Whewell, *History of the Inductive Sciences*, II, 306.

11. James D. Dana, "A New Mineralogical Nomenclature," *Annals of the Lyceum of Natural History of New York*, IV, 1837–1847 (Read: March, 1836), p. 9.

12. *Ibid.*, p. 10.

13. "Review of Aiken, 'A Manual of Mineralogy,' " *NA Rev.*, V (1817), 75.

14. "Gorham's Chemistry," 117.

15. Benjamin Silliman, Jr., "Review of Alger's *Phillips Mineralogy* and Shepard's *Treatise on Mineralogy*," *AJS*, XLVII (1844), 347.

16. C. U. Shepard, "Reply to a Notice of Shepard's Mineralogy," *AJS*, XLVIII (1845), 169.

17. Silliman, "Review of . . . Shepard," p. 348.

18. Whewell, *History*, II, 339–40.

19. Robert Hare, "Encomiums upon . . . Berzelius," 64.

20. *Ibid.*, 67.   21. Whewell, *History*, II, 313.

22. Asa Gray, "Review of John Lindley, 'A Natural System of Botany'," *AJS*, XXXII (1837), 300. In the 1845 edition of his work, Lindley listed 8,935 genera and 92,930 species. See p. 800 of his work for a tabulation.

23. The number is reported in Julius Von Sachs, *History of Botany, 1530–1860*, translated from the German by Henry E. F. Garnsey and revised by Isaac Bayley Balfour (Oxford, 1890), p. 144.

24. T. W. Harris, "Upon the Economy of Some American Species of Hispa," *Boston Journal of Natural History*, I (1835), 141.

25. Torrey's treatise on the plants collected by the Long expedition was published in the *Annals of the New York Lyceum of Natural History*, II (1828), 161; although it had been read two years earlier. With Asa Gray he published *A Flora of North America Arranged According to the Natural System* (2 Vols., 1838–1843). L. C. Beck's *Botany of the Northern and Middle States* (1833), was the next work after that of Torrey to be based upon the natural system.

26. Leonard Jenyns, "Report on the Recent Progress and Present State of Zoology," *Fourth Report of the British Association for the Advancement of Science*, 1834, p. 144.

27. [Dr. Ray], "De Candolle's Botany," *NA Rev.*, XXXVIII (1834), 39.

28. Gray, "Natural System," 294.

29. Sachs, *History of Botany*, p. 116.

30. [Dr. Ray], "De Candolle's Botany," 39; Jenyns, "Report," 146.

31. Godman, *American Natural History*, p. xiii.

32. Gray, "Natural System," 297. Systematists have largely given up the search for a natural system in the same sense that the term was used in the early nineteenth century. Most would probably agree that a system should be called natural if the defining properties of the classes allow us to make more inferences than the defining properties of any other classification system. Beyond this, however, there are various op-

posing points of view. "Community of descent" is considered the main basis for classification by G. G. Simpson: *Principles of Animal Taxonomy* (New York, 1961), p. 53, although he makes certain careful qualifications to his bare statement. The so-called empirical taxonomists, however, hold that the idea of affinity as over-all similarity is logically prior to that of evolutionary affinity. For a sample of this argument, see A. J. Cain and G. A. Harrison, "An Analysis of the Taxonomists' Judgment of Affinity," *Proceedings of the Zoological Society of London,* CXXXI (1958), 85–98. J. R. Gregg, in *The Language of Taxonomy* (New York, 1954), has attempted to apply set theory and symbolic logic to classification, but with equivocal success.

VI. The Limits of Baconianism: History and the Imponderables

1. James D. Whelpley, "Second Letter on Philosophical Analogy," *AJS,* 2nd series, V (1848), 329. At the time he wrote the article, Whelpley was editor of the *American Whig Review.*

2. The qualification "in any unsophisticated sense" is made because it is true that all science does *rest* upon direct observation of phenomena and it is not now considered particularly useful to make an invidious distinction between direct observation and that which rests upon it—at least concerning questions of existence. Even our belief in electrons can be shown, by an extremely long chain of reasoning, to have a direct observational basis; since it has this basis, the existence of electrons is as certain as anything we directly observe. In short, the fact that one cannot see an entity is not considered to pose a particularly important problem today; however, lacking such sophisticated explanations, it was considered quite important a century ago. On this point, see Philipp Frank, *Philosophy of Science: The Link Between Science and Philosophy* (Englewood Cliffs, New Jersey, 1957), pp. 219 ff.; or Gustav Bergmann, *Philosophy of Science* (Madison, 1957), pp. 13–14. P. W. Bridgman reduces all concepts to operational definitions, apparently of equal validity. See particularly the first chapter of *The Logic of Modern Physics* (New York, 1960).

3. No distinction was made at that time between atoms and molecules.

4. [Benjamin Silliman], "Hayden's Geological Essays," *AJS,* III (1821), 56. See also Levi Hedge, *Elements of Logick; or a Summary of the General Principles and Different Modes of Reasoning* (Boston, 1838, from the edition of 1821), pp. 78–79.

5. [F. C. Gray], "Systems of Geology," *NA Rev.,* VIII (March, 1819), 403.

6. Hedge, *Logick,* pp. 97–98.

7. The term "theory" is now generally reserved for a higher, or more complex statement, usually containing several laws, and also definitions. In this connection, see fn. 40.

8. Hans Reichenbach, in *The Rise of Scientific Philosophy* (Berkeley, 1951), p. 230 ff., commenting on the case of the swans, noted that there was another, equally valid induction that should have warned logicians against making the generalization "all swans are white," even before the discovery of black swans in Australia. The rival generalization would have been that color varies among the individuals of the same species. This is essentially the same point as that made by John Stuart Mill, *A System of Logic, Ratiocinative and Inductive* (New York, 1874, from the fourth edition), p. 232. Gustav Bergmann, in a private comment, held that whether or not the black Australian "things" were swans in reality depended upon whether "whiteness" was to be among the defining characteristics of "swan."

9. Thomas Brown, *Lectures on the Philosophy of the Human Mind* (3 vols., Andover, 1822), I, 124.

10. *Ibid.*, 127.

11. "Transactions of the Physico-Medical Society of New York," (review) *American Monthly Magazine and Critical Review*, II (January, 1818), 183. "Hypothesis" was used in approximately the same manner that "theory" is now used in popular terminology when the speaker wishes to distinguish invidiously between "mere theory" and "practice"; the meaning always being that the practice is true and the theory is not. "Hypothesis," on the other hand, is apparently considered more respectable in popular terminology; perhaps because of the widespread confusion of "hypothesis" and "hypothetical," along with the vague knowledge that a hypothesis is something to be tested and not necessarily even claimed to be true. This makes any "hypothesis" an innocuous thing in the popular mind.

12. "Gorham's Chemistry," *NA Rev.*, IX (1819), 124.

13. William James McNevin, *Exposition of the Atomic Theory of Chemistry and the Doctrine of Definite Proportions*, as quoted in the *American Medical Recorder*, III (1820), 64, in a review by T. M. Harris. McNevin (Doctor of Physic; Vienna, 1784. M.D., Columbia College) was co-editor of the *New York Medical and Philosophical Journal and Review*, and professor of chemistry and *materia medica* at the College of Physicians and Surgeons of the University of New York.

14. Eli W. Blake, "Remarks on the Theory of the Resistance of Fluids," *AJS*, XXIX (1836), 274. It is now known that any acceptable form of inductive inference can be reduced to induction by enumeration. In the early nineteenth century, however, this was not known. See, for example, Mill, *Logic*, p. 227. However, being reducible to induction by enumeration and being a direct result of induction by enumeration in the sense that was then intended are different matters; the first is a purely logical concept, while the second involves psychological factors.

15. Whether theories are built upon by "induction by enumeration" or "induction by intuition" has been an issue at least since the classic argument between John Stuart Mill and William Whewell. For their argu-

ment, see Mill, "On Whewell's Philosophy," *Philosophical Review,* LX (1851); Whewell, *Of Induction, With Especial Reference to Mr. J. S. Mill's System of Logic* (London, 1849). Hans Reichenbach brings the two positions together by making a careful distinction between the "context of discovery" and the "context of justification"; the first bringing in psychological considerations and the second being purely structural. *Rise of Scientific Philosophy,* p. 230. Note, however, that Mill did not use "induction by enumeration" in its modern signification. Although he can be attributed a belief in what we now mean by the phrase, in its Baconian sense (*inductio per enumerationem simplicem*) the idea of experiment was specifically excluded. *Logic,* p. 227.

16. In accordance with early nineteenth-century usage, when "imponderable" is used as an adjective, as with "matter," "substance," etc., it means "absolutely devoid of weight." However, when used as a noun, it carries no judgment about weight, but is merely a general name applied to a class of agents.

17. William Whewell, "Demonstration That All Matter is Heavy," *AJS,* XLII (1842), 268.

18. William Whewell, *History of the Inductive Sciences,* Vol. II, 213.

19. Benjamin Silliman, *Elements of Chemistry in the Order of the Lectures Given in Yale College* (New Haven, 1830), Vol. I, 4.

20. Denison Olmsted, *A Compendium of Natural Philosophy Adapted to the Use of the General Reader and of Schools and Academies* (New Haven, 1850; reprint of 1844 edition), p. 249. For a sketch of Olmsted, see Appendix I.

21. Robert Hare, "Letter in Opposition to the Conjecture That Heat May Be Motion and in Favor of the Existence of a Material Cause of Calorific Repulsion," *AJS,* IV (1822), 148.

22. Robert Hare, *A Compendium of the Course of Chemical Instruction in the Medical Department of the University of Pennsylvania* (Philadelphia, 3rd ed., 1836), p. 2.

23. Robert Hare, "Remarks on a Recent Speculation by Faraday on Electric Conduction and the Nature of Matter," *AJS,* XLVIII (1845), 248. In 1844 Faraday had expressed his preference for Boscovich's point-atoms, for "mere centres of force, not particles of matter." This is what Hare referred to in the above quotation as a viewpoint with which he could not concur. For Faraday's viewpoint on the atomic theory, see Lancelot Law Whyte, *Essay on Atomism: From Democritus to 1960* (Middletown, Connecticut, 1961), p. 62. I have previously noted that the Scottish philosophy allowed the transformation of problems of science into problems of the psychology of perception. This is not, however, the same as identifying the world with the perception of it, for Scottish philosophers of common sense clung tenaciously to the belief that there was a real, objective world which one perceived. For them, the argument was that through the grace of God, our senses were in accord with

this world. In other words, theirs was a parallelistic viewpoint rather than a subjective one.

24. [John Ware], "Gorham's Chemistry," *NA Rev.*, IV (1819), 124. He did not, however, accept the hypothesis of Davy. Ware (M.D., Harvard, 1816) was editor at different times of both the *New England Journal of Medicine and Surgery* (1824–1827) and the *Boston Medical and Surgical Journal* (1828). The highest academic post he attained was that of Hersey Professor of the Theory of Physic in the Harvard Medical School, in which chair he was the successor of James Jackson.

25. Hare, *Compendium*, p. 2.   26. Olmsted, *Compendium*, p. 11.

27. Denison Olmsted, "Remarks on Dr. Hare's Essay on the Question, Whether Heat Can Be Ascribed to Motion?" *AJS*, XII (1827), 362.

28. Olmsted, *Compendium*, p. 10.

29. The physical and the chemical viewpoints were coming into opposition on a number of subjects at that time, and if one were writing a history of the development of the modern logic of science, he should probably concentrate on the developments in chemistry in the early nineteenth century. The following quotation from Hare's *Compendium*, p. 1, brings out one of the points of conflict and is, at the same time, the earliest clear statement that I have seen in an American scientific text of the proposition that all the variables of a system interact: "In mechanics, action is said to produce reaction, but in the case of an innate property which mutually causes different portions of matter to be self-attractive, or repellent, it is impossible to distinguish the agent from the reagent. From our first acquaintance with any bodies so situated, they may be said to mutually react, or to exercise reaction." Hare generalized the concept to the point of considering mechanical or physiological systems as being interacting systems of particles or masses (p. 2). A further discussion of the conflict between the physical and chemical viewpoints is in a later chapter.

30. Robert Hare, "A Letter to William Whewell, in Reply to Certain Allegations and Arguments Advanced in a Pamphlet Entitled a Demonstration That All Matter is Heavy," *AJS*, XLIII (1842), 260.

31. Lewis C. Beck, *A Manual of Chemistry* (Albany, 2nd ed., 1834), p. 35. It is not true that it took the development of thermodynamics for men to realize that heat flows from hot to cold until an equilibrium of temperature is reached. This much of the second law of thermodynamics had been a matter of common knowledge for years. See, for example, Nathaniel Chapman, "Thoughts on Animal Heat," *Philadelphia Journal of the Medical and Physical Sciences*, VII (1823), 28. The second law of thermodynamics simply restates the fact in a different framework which allows more implications of the statement to be seen. According to S. Lilley, in terms of what was known in the first quarter of the nineteenth century the material theory of heat was a great deal more reasonable, and more useful, than the mechanical theory, although after

about 1810 a progressive breakdown of its superiority began. "Attitudes to the Nature of Heat about the Beginning of the Nineteenth Century," *Archives Internationales d'Histoire Des Sciences,* I (July, 1948), 630–39.

32. Beck, *Manual of Chemistry,* p. 35.   33. *Ibid.,* p. 32.

34. *Ibid.,* pp. 93–94.

35. Lardner Vanuxem, *An Essay on the Ultimate Principles of Chemistry, Natural Philosophy, and Physiology, Deduced from the Distribution of Matter into Two Classes or Kinds, and From Other Sources* (Philadelphia, 1827), p. iv. "Preliminary" is in quotes because Vanuxem never got around to publishing the more detailed version he promised.

36. *Ibid.,* p. 35.   37. *Ibid.,* pp. 86, 88–91.

38. "Remarks on the Characters and Classification of Certain American Rock Formations," *AJS,* XVI (1829), 254.

39. This is, in essence, the charge made by the reviewer of Vanuxem's *Essay* in the *American Journal of Medical Science,* I (1827), 397.

40. I do not intend to imply that there is now universal agreement on the nature of a theory, for the opposite is more nearly the case. For P. W. Bridgman, theories are apparently of no consequence; science consists merely of reducing unfamiliar phenomena to familiar terms by means of operational definitions: *The Logic of Modern Physics* (New York, 1960; first published, 1927), especially pp. 37–38. Gustav Bergmann, on the other hand, considers theories to be composed of laws (synthetic all-statements) and definitions (which are analytic) connected with each other: *Philosophy of Science,* pp. 31–32. Differing from both of these, Norman Campbell, in *What is Science* (New York, 1952, first published 1921), pp. 85–86, sharply distinguishes between theories and laws. Philipp Frank points out that the best theory is not even necessarily the one that most nearly coincides with observed fact. His point is that the only way to achieve complete agreement is to record all the observations; however, no one would consider such a record an acceptable theory. A theory must be simpler and shorter than the record of observations; hence the acceptance of a theory is always the result of a compromise between the requirement of "agreement with facts" and "simplicity." *Philosophy of Science,* pp. 348–54. Richard Von Mises likewise de-emphasizes the "truth" of a theory and emphasizes the importance of "economy of thought." *Positivism: An Essay in Human Understanding,* translated by Jeremy Bernstein and Roger Newton (Cambridge, 1951), p. 173. The important point is that no modern writer with whose works I am familiar now considers the question of whether theories are "true" or "false" in the same sense that statements of individual fact are true or false to be a very significant question.

41. Joseph Henry, "On the Theory of the So-Called Imponderables," *AAAS Proc.,* VI (1851), 84–91. In the last direct statement I could find on the subject of theory by Vanuxem, he was not willing to go

so far as Hare and Beck in discounting truth or falsity. In the same year he wrote the *Essay*, he argued that even though the igneous theory of the origin of primitive rocks was adequate to explain the facts in question, still it was founded on analogy and ought not to be put in competition with the view or theory which explains the facts, "by causes deduced from positive knowledge." "Proofs, Drawn from Geology, of the Abstraction of Nitrogen from the Atmosphere, By Organization," *AJS*, XII (1827), 91. However, in the *Essay*, p. 50, appears this statement: "The following facts constitute the science of electricity, and whatever principle will explain the whole of these facts, that principle will be its theory; or in other words, upon it the phenomena of electricity will depend." It seems, as one should expect of people in such a transitional period, that he was never quite able to make up his mind on the subject of theory.

42. I could find no evidence that Whewell entertained the concept of closed systems; it is an idea, in fact, that is quite alien to the spirit of Baconianism and to the theological interpretation of nature of the time. A number of chemists, like Beck and Hare, seemed to assume that there were closed systems in nature, but the concept was never formulated by them, or by any other scientists of the period whose works I have seen. Hare did argue, against Whewell, that there was no reason why the inapplicability of gravitation in the case of imponderable bodies should render its employment futile in the case of ponderable bodies. "Letter to William Whewell," 262.

43. Beck, *Manual of Chemistry*, p. 73. Beck's lack of originality in this formulation should not obscure the boldness in making such a claim. Among others, Aepinus, a German electrician, had also saved the one-fluid theory by declaring unelectrified matter self-repellent.

44. Olmsted, *Compendium*, p. 252. Although I follow usage in labeling this "Newton's canon," it can actually be traced at least as far as William of Occam, and it seems to have been a part of the commonly accepted body of scientific thought long before either Newton or Bacon. Edward Wright, for example, in his prefatory epistle to Gilbert's *De Magnete* (1600), used practically the same language in stating this canon that Newton was later to use. Gilbert's work is now readily available in Derek J. Price (ed.), *On the Magnet* (New York, 1958).

45. Beck, *A Manual of Chemistry*, p. 94.

46. Olmsted, *Compendium of Natural Philosophy*, p. 251.

47. See Appendix I.

48. For Silliman's account of his religious differences with Hare, see his letter to Hare of 1857, quoted in Edgar F. Smith, *The Life of Robert Hare An American Chemist* (Philadelphia, 1917), pp. 500–503.

49. Robert Hare, *Experimental Investigation of the Spirit Manifestations Demonstrating the Existence of Spirits and Their Communion with Mortals. Doctrine of the Spirit World Respecting Heaven, Hell, Moral-*

ity, and God. Also, the Influence of Scripture on the Morals of Christians (New York, 1855), pp. 125–26.

50. On this point, see the article by T. A. Conrad, "Notes on American Geology," *AJS*, XXXIII (1839), 237 ff.; and especially the note by Silliman beginning on page 251.

51. John T. Merz in *History of European Thought in the Nineteenth Century* (Edinburgh and London, 1903) makes the point that with the generalization of Schwann in 1839 that "there is one universal principle of development for the elementary parts of organisms however different, and that this principle is the formation of cells," the morphological approach had been exhausted. The fundamental unity of the organization of living beings had been proved and the next problem was to explain their actual diversity. This, he suggests, led directly into the genetic view of nature. Vol. II, 266.

52. [Edward Hitchcock], "Review of 'Outlines of Geology of England and Wales,' by Rev. W. D. Conybeare and William Phillips," *AJS*, VII (1824), 205, 208, 237.

53. [Edward Hitchcock], "The New Theory of the Earth," *NA Rev.*, XXVIII (April, 1829), 266, 291, 260–70.

54. Mark Hopkins, "On Mystery," *AJS*, XIII (1828), 222.

55. On this point, see Kuhn, *The Structure of Scientific Revolutions*, p. 78.

VII. Finalism, Positivism, and Scientific Explanation

1. Samuel L. Mitchill, "Preface" to the *American Edition of Erasmus Darwin, Zoonomia; or the Laws of Organic Life* (New York, 1796), p. xxii. Mitchill was speaking of John Brown (1735–1788), founder of the Brunonian system of medicine.

2. Richard H. Shryock, *Medicine and Society in America*, p. 65.

3. Thomas Jefferson to Caspar Wistar, June 21, 1807. I am indebted to Courtney R. Hall for calling my attention to this and the following letter of Jefferson, in his article, "Jefferson on the Medical Theory and Practice of His Day," *Bulletin of the History of Medicine*, XXXI (1957), 235.

4. Thomas Jefferson to Benjamin Rush, March 6, 1813.

5. Jefferson to Wistar.

6. "On Reasoning in Medicine," *Philadelphia Journal of the Medical and Physical Sciences*, I (1820), 219. Maclurg was one of those who retained a belief in theorizing in spite of the rampant empiricism of his time. See also John D. Godman, "Some Observations on the Propriety of Explaining the Actions of the Animal Economy by the Assistance of the Physical Sciences," *ibid.*, III (1821), 54, for a similar statement; or William H. Shaw, "On the Connexion of Other Departments of Science with

Medicine, Embracing an Investigation of Their Influence on the Existing Doctrines in Regard to the Modus Operandi of Medicines," *ibid.*, XIII (1826), 273. Shaw also saw the need for theory, but was opposed to chemical reasoning in physiology.

7. J. Augustine Smith, "A Lecture on Medical Philosophy," *New York Medical and Physical Journal*, VII (1828), 174, 168, 185. Smith (William and Mary, 1800), had been the President of William and Mary College from 1814 to 1826; and from 1809 to 1811 had been the editor of the *New York Medical and Philosophical Journal and Review*. At the time he delivered the lecture referred to he was lecturer on anatomy in the College of Physicians and Surgeons of the University of New York.

8. *AJS*, I (1818), 416.

9. A., "Geology at Variance with Scripture," *The United States Catholic Magazine and Monthly Review*, V (1846), 309.

10. III (1818), 177. See the quotation above from Jefferson for the same expression, "ingenious dreams," which was a favorite with many writers when they were discussing theories.

11. VI (1818), 416.

12. In the terminology of Levi Hedge this would have been "circumstantial evidence," and would have been entitled to rank as a "presumption." In order to rank as a "violent Presumption," the general currents' existence at that time and in that place would have to be independently known. Thus, the veracity of Moses was a great deal more than a purely religious question, for if his testimony concerning the universal deluge was accepted as authentic, a "violent presumption" could have been established for explanations alleging the activity of water in accomplishing any works which water was known to be able to do. Without such testimony, the general current would have been merely "postulated for the sake of the hypothesis," and this was considered clearly inadmissible. For a further account of Hedge in connection with the issue of theory, see Chapter 6. Note, however, that Hedge's criteria for the admission of explanatory hypotheses was a great deal more liberal than many would have allowed. Rafinesque, for example, in the quotation above, would give no recognition to "probabilities" of this nature.

13. *AJS*, III (1821), 56. For a similar review of Hayden, see *NA Rev*, XII (1821), 134. For a summary of Hayden's book, see E. D. Merrill, *The First One Hundred Years of American Geology* (New Haven, 1924), pp. 81–86. Hayden (1769–1844) was one of the pioneers of American dentistry. He was active in founding the Baltimore College of Dental Surgery and in 1840 he became the first president of the college and professor of the principles of dental science.

14. "Geology," *American Quarterly Review*, VI (1829), 73–104.

15. Olmsted, *Compendium*, p. 266. Olmsted was not exaggerating. For a few examples of the kind of speculations he was talking about, see: G. Gibbs, "On the Connexion Between Magnetism and Light," *AJS*, I (1818), 90, who attributed the aurora borealis to emission of magnetic

rays with electric rays, both of which he, in turn, attributed to light; Robert Hare, who, in "On the Causes of the Tornado, or Water Spout," *APS, Trans.*, n.s., V (1837), 375, developed an electrical theory of storms; or Jason Swaim, "Electro-Meteorological Observation," *AJS*, XXXII (1837), 304, who attributed epidemics to electricity. For the claims made for electricity by medical quacks, see James Harvey Young, *Toadstool Millionaires: A Social History of Patent Medicines in America before Federal Regulation* (Princeton, 1961), especially Chapter II. Olmsted may or may not have known of the work of Du Bois-Reymond, but speculations identifying galvanic electricity and the nervous influence were quite common. On this point, see William Whewell, *History of the Inductive Sciences*, Vol. II, 65 ff., for quotes and citations.

16. I use "ideological" in the sense of "ideological statement" as defined by Gustav Bergmann in *The Metaphysics of Logical Positivism* (New York, 1954), pp. 310 ff. An ideological statement is a statement of value which is mistaken for a statement of fact, and which, accordingly, enters into reasoning on the same level as factual statements. This usage differs markedly from the definition offered by Karl Mannheim, for whom all statements are "ideological," and who connects ideology to a class bias. On this point, see *Ideology and Utopia* (New York, n.d.), p. 57. Although I shall not have much occasion to use the word "ideology" hereafter, it should become clear that the relationship between vitalism and positivism was ideological.

17. Dennison Olmsted, "Thoughts on the Sentiment That the World Was Made for Man," *New Englander*, VII (1849), 17–46.

18. "Positivistic" is here used in a general sense to refer to those who insisted that science must be bound to positive knowledge (about the succession of phenomena in time) and could not hope to attain to "real" causes, in the sense of reason or purpose. The word was not current in that usage in the early nineteenth century, but has since been applied to that viewpoint; it was, however, implied in phrases like "positive knowledge," "positive science," and others, which were in common usage. While the viewpoint was a forerunner of twentieth-century positivism, the two are, nevertheless, very different and should not be confused. Concerning "vitalism" and "finalism," while they are not logically the same, there is such a close connection between them that all those who used vitalism as an explanatory principle (see footnote 57) were also finalists—in fact, vitalism can be considered the form that finalism took in biology. I do not, therefore, find it useful to distinguish between the two viewpoints to the extent of considering them separately, for the same general attitude toward nature is at the foundation of both.

19. It is with great reluctance that I use a term so laden with contemporary meanings as "mechanistic." The word and its derivatives have various and often contradictory meanings in contemporary usage, and none of the common meanings exactly conveys the early nineteenth-century meaning. However, there seems to be no escape from using the

word, for I can think of none less objectionable. In the early nineteenth century, "mechanical philosophy," "mechanism," and sometimes "natural philosophy" were used as synonyms. Olmsted's distinction between "natural philosophy" and chemistry will, therefore, serve quite well to show what I mean by the terms. "In Natural Philosophy, we mean by particles, the *smallest parts* into which a body may be supposed to be divided by mechanical means, without any reference to the different elements of which such particles may be composed. Inquiries of the latter kind belong to chemistry. . ." *Compendium*, p. 10. See also Hare, *Compendium*, p. 2, and A. P. Fourcroy, *Systeme des Connaissances Chimiques* (Paris, 1801), Vol. I, p. xliii. The chemical attitude, of course, taken to its logical extreme, as with John Stuart Mill, results in emergentism. But emergentism was not the issue in the 1820s; on the contrary, it seemed to most people that chemistry was threatening emergents that had already been given, such as life and mind. For Mill's chemism, see Howard C. Warren, *A History of the Association Psychology* (New York, 1921), Chapter IV. For a glossary of contemporary usages of "mechanistic," see Bergmann, *Philosophy of Science,* p. 107. Owsei Temkin, in "Basic Science, Medicine and the Romantic Era," *Bulletin of the History of Medicine*, XXXVII, No. 2 (March, April, 1963), pp. 128–29, has recognized that chemistry was the great threat to vitalism and he has also commented on the connection between empiricism and vitalism.

20. Mitchill, *Zoonomia*, p. xxviii. A famous case of the same nature is Erasmus' translation of a classical refutation of ancient skepticism, which for the first time made Renaissance thinkers acquainted with Pyrrhonism—and in due time a great number adopted it. As Stow Persons, in a private comment, once remarked, when the Mathers became disturbed about Sadducism there was not a single Sadducee in America, but they began to appear shortly after. The same was true of Jonathan Edwards' denunciation of Arminianism. Perhaps there is a peculiar reverse persuasiveness about bad arguments against a position. Less facetiously, it is probably a fact that because of ideological commitments of one kind or another men cannot see some of the logical implications of their ideas until they are pointed out to them by an unfriendly critic. Although of course I cannot document it, I would suggest that these considerations might help explain the peculiar fact that in the early years of the nineteenth century men spent a great deal of time denouncing a position that apparently no one held until somewhat later.

21. See, for example, Olmsted, *Compendium*, p. 2; or Elias Loomis, *Elements of Natural Philosophy* (New York, 1858), p. 13. Taken literally, it would seem that the early nineteenth-century definitions of a law and the usual contemporary definitions are the same. The important distinction, however, is that the present concept of law includes the idea of explanation by subsumption under a higher law—a process that Olmsted had explicitly dismissed as a tautology not worthy of consideration (see Chapter VI).

22. Tayler Lewis, *Six Days of Creation: or, The Scriptural Cosmology, with the Ancient Idea of Time-Worlds, in Distinction from Worlds in Space* (Schenectady, 1855), p. 404.

23. James D. Dana, "Science and the Bible: A Review of 'The Six Days of Creation' of Prof. Tayler Lewis," *Bibliotheca Sacra*, XIII (1856), 80–129, 631–56; XIV (1857), 388–413, 461–524. For a sketch of Dana, see Appendix I.

24. Other parts of the work, however, would certainly have been bitterly opposed; for Lewis' attack on science was also intended as an attack on natural theology.

25. Mark Hopkins, "On Mystery," *AJS*, XIII (1828), 222.

26. Herbert S. Klickstein, "A Short History of the Professorship of Chemistry of the University of Pennsylvania School of Medicine, 1765–1847," *Bulletin of the History of Medicine*, XXVII (1953), 5.

27. (New York, 1822), p. 32. For a sketch of Emmet, see Appendix I. To my knowledge, the next American work which explicitly denied any function to a "vital force" was John W. Draper's *Treatise on the Forces Which Produce the Organization of Plants* (New York, 1844). Although the religious objection still existed when Draper published his work, it was a great deal milder. See, for example, the review in *Princeton Review* XVII (April, 1845), 347.

28. V (1822), 536. The University of New York was a humoralist stronghold, although other graduates did not go so far as Emmet. In the same year that Emmet was awarded his degree, the same institution granted an M.D. to Jacob Dyckman for *An Inaurgural Dissertation on the Pathology of the Human Fluids*. A bitter review, written by either Chapman or Caldwell, was published in the *Philadelphia Journal of the Medical and Physical Sciences*, IV (1822), 344.

29. "Reduction" means, briefly, that the laws and theories of one science can be deduced from the laws and theories of another. This means that before reduction can be attempted one must have a comprehensive theory for both sciences in question. Thus, an argument over reduction was, in a purely logical sense, a bit premature in the early nineteenth century, as it is now. The argument, insofar as it involved the reduction issue, was really over future prospects, not existing possibilities. Reduction, with all its ramifications, is a great deal more complicated than I have made it appear in this brief note, but I think it will suffice for present purposes. For a better discussion, see Bergmann, *Philosophy of Science*, pp. 162–71.

30. Nathan R. Smith, *A Physiological Essay on Digestion* (New York, 1825), quoted in *American Medical Review and Analytical Journal*, I (1825), 504–5, 513. The *Journal*, which Smith co-edited with Eberle, contained quotations outlining the entire argument of the work. In addition to editing the *Journal*, Smith was at that time professor of anatomy and physiology at the University of Vermont. Smith did use the language of cause and effect, but note the last quotation, where it be-

comes obvious that he meant "succession in time" or "proximate cause."

31. D. Francis Condie, "On Digestion," *Philadelphia Journal of the Medical and Physical Sciences*, X (1825), 359, 364. Eberle and Chapman were arch-rivals in science. The first issue of Eberle's *American Medical Review* was devoted largely to a critical examination of articles in Chapman's *Journal*. Condie (M.D., University of Pennsylvania, 1818) was known chiefly as a writer of medical texts; his special field of interest was diseases of women and children.

32. "Observations and Reasoning in Medicine," lecture of 1791 in Dagobert D. Runes (ed.), *The Selected Writings of Benjamin Rush* (New York, 1957), p. 246. Rush added that he also believed that no empiric ever gave a medicine without cherishing a theoretical indication of cure in his mind.

33. "Thoughts on Animal Heat," *Philadelphia Journal of the Medical and Physical Sciences*, VII (1823), 23.

34. Modern vitalism still rests its case on unanswered questions which it somehow makes appear unanswerable. See, for example, Bruno Kisch, "What Keeps Men Alive?" in Chandler McC. Brooks and Paul F. Cranefield, *The Historical Development of Physiological Thought* (New York, 1959), pp. 309–34. In the same volume there is a short historical sketch of vitalism by Iago Galdston, pp. 291–308. For a more detailed history of vitalism, see Hans Driesch, *Geschichte des Vitalismus* (Leipzig, 2nd ed., 1922). Many contemporary vitalists deny that the term is applicable to them, and they characteristically cloak their opinions in the language of cybernetics. The leading spirit in this trend has been Ludwig von Bertalanffy, whose *Modern Theories of Development* (J. H. Woodger, trans., Cambridge, 1933), is still the best statement of the viewpoint (although it antedated the word "cybernetics"). *The General Systems Yearbook*, edited by Bertalanffy and A. Rapoport beginning in 1956, is the chief source for information concerning recent developments.

35. Klickstein, "Professorship of Chemistry," p. 64.

36. *A Discourse on the Connexion between Chemistry and Medicine* (Philadelphia, 1818), p. v.

37. N. Chapman, "On the *Modus Operandi* of Medicines," *Philadelphia Journal of the Medical and Physical Sciences*, II (1821), 295.

38. "Remarks on Dr. Hare's Essay on the Question, Whether Heat Can Be Ascribed to Motion?" *AJS*, XII (1827), 368.

39. This tendency to impose arbitrary limits on the range of empirical science was Asa Gray's chief complaint against Agassiz's view of species. For a good account of this controversy, see Dupree, *Asa Gray*, Chapter XII.

40. *Zoonomia*, pp. xxviii–xxix. Chevreul, in his work on fats (1811–1823), had found that the same principles and combinations held in the animal world as in the vegetable; this was perhaps the highest achievement of organic chemistry up to that time. One of the characteristics of

all organic products was understood to be the fact that "they cannot be formed by the direct union of their principles." See L. C. Beck, *A Manual of Chemistry* (Albany, 2nd ed., 1834), p. 362. Although it is often said that Wohler in 1828 performed the first synthesis of a substance in a living organism, this is not strictly true. Wohler formed urea from ammonium cyanate, which could not then be formed from its elements.

41. "Thoughts on Animal Heat," *Philadelphia Journal of the Medical and Physical Sciences*, VII (1823), 23.

42. "*Modus Operandi*," pp. 300–302. See also, Condie, "On Digestion," pp. 360–64, for a summary of experiments which were taken to refute the principle of chemical action. There were also legitimate objections that could be made to all the experiments allegedly proving the chemical thesis. On this point, see Henry Bond, "On the Production of Animal Heat," *Philadelphia Journal of the Medical and Physical Sciences*, X (1825), 307 ff. The development of the achromatic microscope and the generalization of Schleiden and Schwann later gave an added boost to vitalism, and even threatened to recall the Platonic theory of ideas to scientific respectability. If all primordial cells were the same, it was argued, how could one explain the fact that developed organisms were different without calling in a peculiar life force which determined the forms of each organism? For a sample of this argument, see W. I. Burnett, "The Relations of Cells to the Physical and Teleological Views of Organization," *AJS*, 2nd series, XV (1853), 87–94. G. J. Goodfield, in *The Growth of Scientific Physiology* (London, 1960), has also argued that the vitalists had a good case in terms of early nineteenth-century knowledge. See especially Chapters III and V.

43. Usher Parsons, "On the Connexion between Cutaneous Diseases Which Are Not Contagious, and the Internal Organs," [Boylston Prize Dissertation, 1830] *Boston Medical and Surgical Journal*, III, Nos. 27 & 28 (Tuesday, August 17, 1830), 430. If the body itself is viewed as a chemical system, the difference in effect is readily explainable from a chemical viewpoint, for the variables of a system interact. In this connection it must be said that whether or not the variables of a system did in fact interact was still unsettled. Berthollet had so argued in his *Essai de Statique Chimique* (Paris, 1803), and it was accepted almost as a maxim by Robert Hare (see Chapter vi). Others, however, were still following Torbern Bergman, whose *Dissertation on Affinity* was translated in London in 1785. Bergman's viewpoint was that elements have a constant "affinity," and in a compound that element having the strongest "affinity" gives the characteristics to the compound. Even Berthollet, however, had more of a mechanical, as the term is being used here, than a chemical viewpoint, for he did not distinguish between mixtures and compounds.

44. William Whewell, *History of the Inductive Sciences*, Vol. II, 489. Somewhat later, Jeffries Wyman's studies of useless organs proved quite damaging to finalism. For an account of his studies and their contribution to evolution theory, see A. S. Packard's "Memoir" in *National Acad-*

emy of Science, Biographical Memoirs, II (1886), 92–93. For a sketch of Wyman, see Appendix I.

45. "Some Observations on the Propriety of Explaining the Actions of the Animal Economy by the Assistance of the Physical Sciences," *Philadelphia Journal of the Medical and Physical Sciences,* III (1821), 46–47.

46. *Ibid.,* pp. 54, 53. See also, Robert Hare, *Compendium,* p. 487, where he suggested that the animal gland might be profitably compared to a Voltaic circuit.

47. "On the Doctrine of Sympathy," *Philadelphia Journal of the Medical and Physical Sciences,* VI (1823), 337. A part of the reason that "one could not help but note the dependence of the different textures and functions on each other" may have been that the taxonomy of the time directed that one not be concerned about differences of structure which were not accompanied by corresponding differences in the rest of the organization. On this point, see Chapter V.

48. "French materialist" was a common term of disapprobation in early nineteenth-century America. An anonymous pamphleteer, in attacking Thomas Jefferson's scientific beliefs, could think of nothing worse to call him than a follower of Voltaire and the French Encyclopedists, "the imps who have inspired all the wickedness with which the world has of late years been infested." "Observations upon Certain Passages in Mr. Jefferson's Notes on Virginia Which Appear to Have a Tendency to Subvert Religion and Establish a False Philosophy." Pamphlet dated 1804 in the collection of the Newberry Library, Chicago, p. 29.

49. Godman, "Sympathy," pp. 348–49.

50. Charles Caldwell, "Thoughts on Sympathy, in A Letter from Charles Caldwell, M.D. to N. Chapman, M.D.," *Philadelphia Journal of the Medical and Physical Sciences,* III (1821), 302, 323.

51. John Eberle, "Observation on 'Thoughts on Sympathy, in A Letter from Charles Caldwell, M.D. to N. Chapman, M.D.'," *American Medical Review,* V (1822), 356.

52. *Ibid.,* p. 354. One of the best illustrations of the ideological connections of science is in the changing position of vitalists on the question of solidism vs. humoralism. In the early eighteenth century humoralism had been equated with vitalism, for the humors were thought to be regulated by "animal spirits," something like a "vital force," while the mechanical explanations of the solidists seemed to do away with these spirits. By the 1820s, however, the humors were thought to be regulated by chemical action, and this seemed to do away with the "vital force" that was alleged as the cause of mechanical activity. Early in the nineteenth century vitalists were therefore solidists, in contrast to their brethren of a century earlier. On the connection between vitalism and humoralism in the eighteenth century, see Mary A. B. Brazier, "The Historical Development of Neurophysiology," in *Handbook of Physiology;* Section I, Vol. I (Washington, 1959), 8–9.

53. Richard Harlan, "On the Generation of Animal Heat," *Philadel-*

phia Journal of the Medical and Physical Sciences, II (1821), 250.

54. S. N., "Claims of Medicine to the Character of Certainty," *Monthly Journal of Medical Literature and American Medical Student's Gazette,* I (1832), 42.

55. If my subject were the history of medical practice rather than the history of scientific thought I would attach a great deal of importance to the fact that vitalism seemed almost inevitably to lead to the notion that diseases and symptoms were the same thing, and that simple symptomatic treatment was indicated in all cases. As the anonymous writer said, the sensibility pervading the system "pointed out the particular organ affected and indicated the kind and degree of action occurring." From this, it follows: (1) that treating the symptoms is the same thing as treating the disease, and (2) that a clinical diagnosis is always more accurate than a post-mortem autopsy. These are only two of the most extreme conclusions to which the principle leads. For an example of a physician who took both these conclusions quite seriously, see Thomas D. Mitchell, "Morbid Anatomy Considered in Its Relation to the Practice of Medicine," *North American Medical and Surgical Journal,* IV (1827), 52. By another chain of reasoning it also leads to the thesis of medical empiricism, which was well-stated in "A Knowledge of Anatomy Does Not Contribute to the Discrimination or Cure of Diseases," *Medical Reformer,* I, No. 2 (February 1, 1823), 25.

56. Samuel Jackson, "On Vitality and the Vital Forces," *Philadelphia Journal of the Medical and Physical Sciences,* XIII (1826), 68–69.

57. In modern terminology, I think we could say that Jackson's argument implies that the existence of a vital principle is a "meaningless" question. G. J. Goodfield, in *The Growth of Scientific Physiology,* p. 75, has distinguished two vitalistic attitudes, "explanatory vitalism" and "descriptive vitalism." While it is also my point that all those who used the language of vitalism did not have the same attitudes toward the nature of life, I wish to emphasize that the significant division was between those who placed some theoretical obstacle between physics-chemistry and life and those who did not. I see no reason to consider this latter group—which includes Godman, Jackson, and Bartlett—as vitalists at all, even though they occasionally used the language of vitalism. With them the terms have been reduced to a convention that is both practically and theoretically empty. In other words, in Goodfield's terminology I am restricting the meaning of "vitalism" to "explanatory vitalism."

58. Bernard's definition was as follows: ". . . la propriété que possède tout élément anatomique (c'est-à-dire le protoplasme qui entre dans sa constitution) d'être mis en activité et de réagir d'une certaine manière sous l'influence des excitants extérieurs." Quoted in L. E. Bayliss, *Principles of General Physiology* (London, 1960), Vol. II, 196. The only addition Bayliss makes to Bernard's definition is to connect irritibility with the requirement, derived from thermodynamics, of cellular maintenance of a steady state by continuous expenditure of energy. p. 193.

59. "On the Vitality of Matter," *AJS*, XV (1829), 54.

60. Footnote, p. 62. Silliman no doubt added the note because he did not wish to leave the vital principle resting on the same shaky foundation on which Galileo's Aristotelian opponents had left the non-existence of the Medicean stars. The writer of the article had argued, in effect, that the "apparent animation" was really a function of the microscope, just as Galileo's opponents had argued that the Medicean stars were in the telescope. Silliman's numerous "editor's notes" to articles in his journal would make a fascinating study in themselves.

61. "On the Affiliation of the Natural Sciences," in *Medical and Physical Researches; or, Original Memoirs in Medicine, Surgery, Physiology, Geology, Zoology and Comparative Anatomy* (Philadelphia, 1835), p. xiii.

62. *An Essay on the Philosophy of Medical Science* (Philadelphia, 1844), p. 194. Being a consistent empiricist, Bartlett did not, however, conclude that chemistry could explain all the activity of the body. For an excellent analysis of Bartlett's book, see Erwin H. Ackerknecht, "Elisha Bartlett and the Philosophy of the Paris Clinical School," *Bulletin of the History of Medicine*, XXIV (1950), 43–60. According to Ackerknecht, this *Essay* is the only systematic formulation of the philosophical approach of the Paris clinical school of Bayle, Laennec, Chomel, and Louis. Bartlett, a practicing physician, had studied under Pierre Louis in Paris.

63. *Ibid.*, pp. 32, 79–80.   64. *Ibid.*, p. 278.

65. So much has been written about theological interpretations of geology that I could add very little by detailing the position. For a good account, see Charles C. Gillispie, *Genesis and Geology: A Study in the Relations of Scientific Thought, Natural Theology, and Social Opinion in Great Britain, 1790–1850* (Cambridge, 1951).

66. [Jared Sparks], "Review of An Abstract of A New Theory of the Formation of the Earth, by Ira Hill, A.M.," *NA Rev.*, XVIII (1824), 279. Sparks (LL.D., Harvard, 1843) at the time of writing the article was editor of the *North American Review*. He is most remembered as a historian, especially for his twelve-volume edition of the *Writings of Washington, With a Life*, the three-volume *Life of Gouverneur Morris*, and his editorship of the *Library of American Biography*, to which he personally contributed eight volumes.

67. [J. G. Cogswell], "Dr. Webster on St. Michael and the Azores," *NA Rev.*, XIV (1822), 48–49.

68. "Observations on the Geology of the United States," APS, *Trans.*, n.s., I (1818), 2–3.

69. John A. DeJong, *American Attitudes Toward Evolution Before Darwin* (Thesis, University of Iowa, 1962), p. 95.

70. "Essay on the Formation of Rocks, or An Inquiry into the Origins of Their Present Form and Structure," *Journal of the Academy of Natural Sciences of Philadelphia*, I (1818), 262–63.

71. [Edward Hitchcock], "The New Theory of the Earth," *NA Rev.,* XXVIII (April, 1829), 286. See also his "On Certain Causes of Geological Change Now in Operation in Massachusetts," *Boston Journal of Natural History,* I (1835), 69.

VIII. The Inductive Process and the Doctrine of Analogy

1. Thomas Brown, *Inquiry into the Relation of Cause and Effect* (Andover, 1822). The syllogism is a summation of the entire argument. See also [Samuel Tyler] "Natural Theology," *Baltimore Literary and Religious Magazine,* VI (July, 1840), 301–24; James Brazer, "Review of the Argument in Support of Natural Religion," *Christian Examiner,* XIX (November, 1835), 137–62; or [Theophilus Parsons], "Tendencies of Modern Science," *NA Rev.,* LXXII (January, 1851), 84–115. Kierkegaard, of course, made the Humian argument an essential element of his theology.

2. Francis Wayland, "Discourse on the Philosophy of Analogy," in Joseph L. Blau (ed.) *American Philosophic Addresses, 1700–1900* (New York, 1946), p. 348. Edwin G. Boring, in *A History of Experimental Psychology* (New York, 2nd ed., 1957), p. 207, makes the point that men have always accepted readily such contentions as that having a mental ability is explained by the fact that one has a faculty for that quality. Such naming he calls "word magic," but he recognizes that it was a major component of the philosophy of Reid and Stewart.

3. *Ibid.,* pp. 350–51.

4. *Ibid.,* pp. 352, 353. Note here that Wayland considers induction to be a simple additive process in the same sense that Tyler did. The generalization, which goes beyond what one has seen, was not a part of induction for Wayland, as it was not for most of his contemporaries.

5. *Ibid.,* p. 354.   6. *Ibid.,* p. 353.

7. To forestall an obvious objection, let me say that I am not falling into the same trap that Mill did when he argued that the uniformity of nature was a sufficient justification for the inductive process. He was, of course, accused of circular reasoning (indeed, he admitted as much himself) in alleging an inductive generalization as the justification for the process of inductive generalization. Whether or not the statement, "Nature is uniform," is an inductive generalization, and whether or not it can be used to justify the inductive process, are two decidedly different questions.

8. "Rules of Reasoning in Philosophy," Rule II, Book III, *Mathematical Principles of Natural Philosophy.* Newton's work is now readily available in Volume XXXIV, *Great Books of the Western World* (Chicago, 1952).

9. But it is a great deal more extreme than that later made by Way-

land himself. By 1854 the only functions he would claim for analogy were (1) to answer the a priori argument, and (2) to provide suggestions for research. See his *The Elements of Intellectual Philosophy* (Boston, 1854), where analogy is given only four pages (337–40).

10. W. P. D. Wightman makes the point that G. G. Stokes was able to develop the theory of fluorescence (1852) only by accepting what seemed a necessary inference even when the inference seemed absurd in the light of contemporary knowledge. See *The Growth of Scientific Ideas* (New Haven, 1953), p. 266. Stokes was able to explain the phenomena of fluorescence ("internal dispersion") by rejecting for a particular instance what Newton had demonstrated to be true for dispersion of light in general. The essence of Stokes' theory (which is now well-established) was that certain substances are capable of absorbing radiation of one wave length and emitting it at a longer wave length. Modern physics, of course, is full of such rejections. On this point, see Albert Einstein and Leopold Infeld, *The Evolution of Physics* (New York, 1961).

11. Benjamin Rush, "Outlines of the Phenomena of Fever," in *Medical Inquiries and Observations* (Philadelphia, 1809, 3rd ed., revised), III, 6–7, 16, 18–21, 36.

12. The earliest example I have been able to find of an attack upon Rush's reasoning, as distinguished from his conclusions, is Elisha Bartlett, *An Essay on the Philosophy of Medical Science* (Philadelphia, 1844), pp. 226–27. For a bitter attack on some of Rush's practices, see "Sangrado's Practice contrasted with Dr. Rush's," *The Medical Reformer*, I, No. 2 (1823), 68.

13. James D. Dana, "On the Analogy Between the Mode of Reproduction in Plants and the 'Alternation of Generations' Observed in some Radiata," *AJS*, 2nd series, X (1850), 341.

14. *AAAS, Proc.,* II (1850), 447. See Asa Gray's work on the same subject, pp. 438–44. About Peirce's paper, Gray later said he "*ran off into the sky*, dementedly"—A. G. to Charles Darwin, June 1863. Chauncey Wright, "The Uses and Origin of the Arrangement of Leaves in Plants," *Philosophical Discussions*, 296–328, at Gray's urging made a mathematical investigation and completely destroyed idealistic notions, substituting a naturalistic explanation derived from the theory of evolution. This was termed by Dupree (*Asa Gray*, p. 291) one of the most important bits of research in "the destruction of the world of Agassiz and the erection of the world of Darwin." *AAAS Proc.,* II (1850), 307. Because of its hypothetical nature, Elias Loomis rejected Kirkwood's analogy. "On Kirkwood's Law of the Rotation of the Primary Planets, *AJS*, 2nd series, XIV (1852), 210. For the reception by American scientists, see *Annual of Scientific Discovery* (1850), pp. 335–38; and for the British Association comments, see *Ibid.* (1851), p. vi.

15. See, for example, John C. Warren, *A Comparative View of the Nervous and Sensorial System in Man and Other Animals* (Boston,

1822). For an argument to the effect that Warren's work did not threaten phrenology, see B. H. Coates, "Comments on Some of the Illustrations Derived by Phrenology from Comparative Anatomy—With Reference to the Late Review of Dr. Warren's Work on the Nervous System," *Philadelphia Journal of the Medical and Physical Sciences*, VII (1823), 58 ff. A number of prominent American scientists, including Benjamin Silliman, accepted phrenology as a valid science. See Silliman's article "Phrenology," in AJS, XXXIX (1840), 65. Silliman's acceptance was based upon an analogy between nerve fibers and blood vessels. On phrenology in America in general, see John D. Davies, *Phrenology, Fad and Science: A Nineteenth Century American Crusade* (New Haven, 1955). The best analysis of phrenology in its relation to other nineteenth-century movements in psychology is in Boring, *A History of Experimental Psychology*, pp. 54–62.

16. Amos Eaton, "The Number Five, The Most Favorite Number of Nature," AJS, XVI (1829), 172–73.

17. Amos Eaton, "Geological Prodromus," AJS, XVII (1830), 63.

18. [G. W. Featherstonehaugh], "Geology," NA Rev., XXXIII (1831), 471. Featherstonehaugh, an English emigre, later became editor of the short-lived *Monthly American Journal of Geology and Natural Science*. He had already, by the time he wrote this review, become a convert to classification of strata by fossil content. Classification of strata by mineral content, not the primacy assigned to water, was considered to be the most important element of Wernerian geology.

19. Samuel Tyler, "On Philosophical Induction," AJS, 2nd series, V (1848), 321. Although Tyler made the point more concisely here, one can gather the same ideas from the earlier articles analyzed in chap. iii. Elisha Bartlett, *Medical Science*, p. 201, quotes a long passage from an address by Oliver Wendell Holmes on the subject of "Democratic Science."

20. James D. Whelpley, "Second Letter on Philosophical Analogy," AJS, 2nd series, V (1848), 329–30. Whelpley did not explain how one moved from the divisible and destructible, by analogy, to the indivisible and indestructible.

21. J[ames] D. W[helpley], "Life and Writings of Coleridge, Chapter II," *American Whig Review*, X (1849), 635.

22. That is, a kind of "induction by substitution" as opposed to the conventional building process from particular to general law and then back to particular. Mill argued that this was the way inductive reasoning actually proceeded, in his *Logic*, pp. 223–24. Bertrand Russell, in *Problems of Philosophy* (New York, 1959, first published 1912), pp. 79–80, argued that on the basis of probability theory, reasoning directly from *many* particulars directly to another particular was preferable to arguing by way of generalizations, for all empirical generalizations are more uncertain than the instances of them. Within the framework of the theory

of probability, however, this becomes quite different from the older "induction by substitution."

23. Whelpley, "Life and Writings of Coleridge," pp. 635–36.

24. Terence Martin, in *The Instructed Vision: Scottish Common Sense Philosophy and the Origins of American Fiction* (Bloomington, Indiana, 1961), has also noted the paradoxical fact that Scottish common-sense philosophy anticipated the very transcendentalist, intuitionist ideas it did its best to avoid. See especially p. 11.

25. Denison Olmsted, "On the Present State of Chemical Science," *AJS*, XI (1826), 352.

26. The quotation is from William Prout, *Chemistry, Meteorology and the Function of Digestion Considered with Reference to Natural Theology* [Bridgewater Treatise] (London, 1834), p. 97, who attributes it to Paley. The tortuous reasoning in the first section of Prout's book, dealing with chemistry, is an admirable illustration of the difficulty natural theologians had with chemistry. For another effort, see George Fownes, *Chemistry, As Exemplifying the Wisdom and Beneficence of God* [Actonian Prize Essay for 1844] (New York, 1844).

27. *Rapports du Physique et du Moral de l'Homme*, I, 299.

28. *History of the Inductive Sciences*, II, 644.

29. Richard Harlan, *Medical and Physical Researches*, pp. xvii, xviii.

30. Thomas McCulloch, Jr., "On the Importance of Habit as a Guide to Accuracy in Systematical Arrangement," *Boston Journal of Natural History*, IV (1844), 406. On the relatively few physiological studies, see Richard H. Shryock, *Medicine and Society in America, 1660–1860* (New York, 1960), p. 129.

31. That is to say, final causes were not in general use *in science*. Those who used the positivistic arguments were finalists, but they insisted that final causes were in another realm from that of science. This was, in fact, the basis of their denigration of science. It is possible that the use of final causes as explanatory elements in science may have been an effort to rescue science from the inferior status it had been assigned. I do not think the fact that Rush provided an outstanding example of analogical thinking in any way endangers the generalization that "analogy was not in such general use in the early part of the century." Rush was unique in a number of ways, and even though a very rudimentary analysis could have shown that his argument was based upon analogy, he did not recognize it to be so grounded. It is the *belief* that analogies are legitimate that I find to be growing after the late 1820s.

32. Reverend Frederick Beasley, *A Search of Truth in the Science of the Human Mind, Part First* (Philadelphia, 1822), pp. 19–20, 329. Part Second was never published. Beasley, the leading defender of John Locke in America, was the Provost of the University of Pennsylvania. Levi Hedge, *Elements of Logick*, p. 87. See below for a discussion of his attempt to distinguish analogy from induction.

33. Brazer, "Review of the Argument in Support of Natural Religion," 138, 140, 145–48.

34. *Ibid.*, 160.

35. "Natural Theology," *Bibliotheca Sacra*, II (May, 1846), 260.

36. [T. Parsons], "Tendencies of Modern Science," *NA Rev.*, LXXII (January, 1851), 114.

37. [John] E[berle], "Review of Combe's Essays on Phrenology (American Edition)," *American Medical Review and Journal*, I (1824), 89, 90.

38. Gray, "Systems of Geology," 403. John P. Harrison, "Observations on Gall and Spurzheim's Theory," *Philadelphia Journal of the Medical and Physical Sciences*, XI (1825), 233. William H. Shaw, "On the Connexion of Other Departments of Science with Medicine, Embracing an Investigation of Their Influence on the Existing Doctrines in Regard to the Modus Operandi of Medicines," *Philadelphia Journal of the Medical and Physical Sciences*, XII (1826), 273.

39. Ebeneezer Emmons, "Circulation in Vegetables," *AJS*, XXVI (1834), 99. Emmons (M.D., Williams College, 1818) was at that time professor of natural history at Williams College. Although best known as a geologist, the originator of the "Taconic Theory," he also did important work in botany. When the Civil War began, he was conducting a geological survey of North Carolina, and was held in the South until his death in 1863.

40. J. H. Steele, "Notice of Snake Hill, and Saratoga Lake and Its Environs," *AJS*, IX (1825), 3.

41. Cuvier's comparative anatomy is brilliantly analyzed by Lloyd G. Stevenson in "Anatomical Reasoning in Physiological Thought," in Chandler McC. Brooks and Paul E. Cranefield (ed.), *The Historical Development of Physiological Thought* (New York, 1959), pp. 27–38. However, Stevenson does not sufficiently recognize the element of good fortune in Owen's success with the bird. The falsity of Cuvier's principle, despite its early success, has been clearly demonstrated in a number of significant cases. The best-known case is that of the chalicotheres, whose feet imply a carnivore but whose skull implies an ungulate. On this point, see George G. Simpson, *Principles of Animal Taxonomy* (New York, 1961), p. 44.

42. Hedge, *Logick*, pp. 26, 88. Note that Hedge employs the same term "analogy" to signify two different processes. In the first instance, it is the act of comparing made in the inductive process; and in the second, it is a "bad" inductive generalization. This confusion was quite common.

43. [Willard Phillips], "Hedge's *Logick*," *NA Rev.*, IV (1816), 90, 92. Phillips accorded the work general high praise and credited Hedge with having written the first text for students containing an extended discussion of the modes of reasoning actually used by scientists—induction and analogy. Phillips (Harvard, 1810) was a lawyer, jurist, and

frequent contributor to review organs. For several years he was editor and publisher of *The American Jurist*.

44. Thomas Reid, *Essays on the Intellectual Powers of Man* (Cambridge, 3rd ed., 1852), pp. 18–19. Reid attempted to distinguish between analogy and induction in the same manner and with much the same success as Hedge.

45. Although the existence of life on other planets was a much debated issue, I have not seen any example from either side of the argument, before Mill, of a recognition that some similarities, or the lack of them, would be more convincing than others. John Stuart Mill, in *A System of Logic,* p. 395, notes that in cases where it has been shown that conditions necessary for life on earth do not exist on another planet, such life as does exist must be dependent upon completely dissimilar causes; therefore, every similarity becomes, not an argument for life on that planet, but an argument against it. Of course, Mill's argument was as specious as that of those who argued against him, but he did note the important point that conditions necessary for the support of life as we know it must be observed if an argument for the existence of life on other planets is to have much weight. Where such conditions do not exist, contrary to Mill, similarities are neither arguments for nor against the existence of life; they are merely irrelevant.

46. This argument is really the reverse of that which was used to prove the existence and character of God by the natural theologians of the eighteenth and nineteenth centuries. In natural theology, God's character was to be deduced from a consideration of his works. In the justification for induction, the interpretation of nature was to depend upon a prior knowledge of God's character—which had, in turn, been derived from Nature. It therefore reduces to the same circular argument which Mill used to demonstrate the validity of induction.

47. Nathan Smith, *A Physiological Essay on Digestion* (New York, 1825). The review referred to is in *The Medical Review and Analytical Journal,* I (1825), 504–25.

48. For a good example of this attitude, see "On the Vitality of Matter," *AJS,* XV (1829), 54, where an anonymous contributor rejects the findings of the microscope because "conclusions drawn from the analysis were unsupported by analogy throughout the visible creation." Dr. John Ware, in a review article, "Hale's Dissertations," *NA Rev.,* XIV (1822), 263 ff., similarly cautions against accepting the results of experiments.

IX. Science, Theology, and Common Sense

1. James R. Manley, "Annual Address Delivered Before the New York State Medical Society, 1827," *American Medical Recorder,* XII (1827), 465.

2. James Maclurg, "On Reasoning in Medicine," *Philadelphia Journal of the Medical and Physical Sciences,* I (1820), 217.

3. Even though the American motives for retaining a mild catastrophism were not primarily scientific, it is not necessarily to their discredit that they refused to rush headlong into the extreme uniformitarianism of Lyell and his British followers. I know of no contemporary geologist who believes that the formation of the earth was as uniform a procedure as Lyell and his followers alleged.

4. Maclurg, "Reasoning in Medicine," p. 229.

5. Suggestions that the imponderables might be modifications of the same force were quite common throughout the period, but they could not lead to any significant research until two conditions were met: (1) a transferable unit, such as "work," was developed to give meaning to the notion; and (2) a great deal more than was known about chemical activity was learned, making it possible to dispense with the "vital principle" and apply the same reasoning to metabolism. The belief that imponderables were modifications of the same force seems to have been one of the elements derived from *Naturphilosophie,* although, of course, no American or British scientist would have admitted such an un-Baconian origin. Kant had held that dissimilars could not interact with each other; the discovery that the imponderables interacted led to the belief that they were in reality the same force. This seems to have been the consideration motivating Ampere's work, and it seems to have had a great deal to do with the disappearance of the idea of "imponderable fluids." The impact of *Naturphilosophie* on nineteenth-century physical science is one of those subjects in desperate need of detailed study.

6. Although I think it fair to say that the generally held opinion now is that "science" is singular and the divisions are only arbitrary, there are still dissenters, of whom F. S. C. Northrop is perhaps the best known. See his *The Logic of the Sciences and the Humanities* (New York, 1947). See also A. D. Ritchie, *Studies in the History and Methods of the Sciences* (Edinburgh, 1958). Their contention is that each science is marked by its peculiar subject matter and by its own "scientific method." In the early nineteenth century, a unique subject matter and a method of its own had to be clearly marked out before a discipline could be admitted on equal grounds with the "other sciences." For this reason, it was believed that mineralogy would no longer be a science in its own right if it used chemical classifications. The argument that it had no subject matter of its own was also used against natural theology's claim to be a science, but in this particular case, as the previous chapter indicated, the argument could be turned around, for the subject matter of natural theology was *all* of nature. This could be taken to mean that it was, in fact, the only fully autonomous science.

Index

Adanson, Michel, 115
Adaptation, 153, 180
Adrian, Robert, 15, 17
Agassiz, Louis, 59, 246
Alternation of generations, 173
American Academy of Arts and Sciences, 13, 173
American Antiquarian Society, 16
American Journal of Science and Arts, 17–18, 39, 56, 71, 88, 97, 110
American Medical Recorder, 148
American Medical Review and Journal, 16, 183
American Philosophical Society, 13, 97, 98, 161
American Quarterly Review, 46, 71
American Whig Review, 177
Ampere, Andre-Marie, 132
Analogy, 167–80, 195; and induction, 72–73, 187; and natural theology, 178–83; and scientific explanation, 174; between animals and plants, 130, 172–73, 184–85, 189; in classification, 173; in geology, 136, 185–186; in mathematical series, 173–74; in medicine, 171–72, 184; in phrenology, 184; in physiology, 180; in vertebrate paleontology, 186
Analyst, 17
Anatomy, comparative, 153, 180, 186, 196, 274
Andover Theological Seminary, 51, 87
Anthropology, 22, 23
Anti-Catholicism, 83
Aristotle, 71
Arminianism, 263
Astronomy, 22, 23, 29
Atheism, 79
Atomic theory, 119–20, 122, 128, 176–77, 254
Avogadro's Law, 122

Bache, A. D., 24, 97, 202 (biog.)
Bachman, John, 203 (biog.)
Bacon, Sir Frances, 63, 71, 82, 103, 118, 179–80, 185, 197, 244–45
Baconianism, 63–85, 100, 177, 192, 198, 259; criticism of, 67–68; influence of, 77; limitations of, 92, 98, 99, 101, 106, 118–19, 138, 166; and the atomic theory, 119–20; and Christianity, 81–82; and classification, 65–66, 72, 86–87, 102; and generalization, 89; and geology, 91, 135, 142–43; and the imponderables, 123–24, 132, 168; and logic, 247; and medicine, 140–41; and the natural system, 114; and psychology, 75–76, 84; and Scottish philosophy, 55, 70–71, 190, 198–99
Bailey, Jacob W., 203–04 (biog.)
Baltimore Literary and Religious Magazine, 71, 83
Baron, George, 16
Bartlett, Elisha, 155, 158–59, 195, 196, 248
Beasley, Frederick, 181, 186, 194
Beaumont, William, 4
Beck, Lewis C., 114, 127–28, 133, 196, 204 (biog.)
Bell, John, 17, 183, 237
Bernard, Claude, 158
Bertalanffy, Ludwig von, 265
Berzelius, Jöns Jakob, 111
Bible, 56, 81
Bibliotheca Sacra, 56
Blake, Eli, 122
Bonaparte, Charles L., 204 (biog.)
Boorstin, Daniel J., 239
Boring, Edwin G., 270
Boston, Society of Natural History, 72
Botany, 22, 23, 29, 112–13
Boyle's Law, 100